Recent Advances in Biotechnology

(Volume 9)

Biotechnology and Drug Development for Targeting Human Diseases

Edited by

Israel Valencia Quiroz
Phytochemistry Laboratory, UBIPRO
Superior Studies Faculty (FES)-Iztacala
National Autonomous University of México (UNAM)
Tlalnepantla de Baz, México State, 54090
Mexico

Recent Advances in Biotechnology

(Volume 9)

Biotechnology and Drug Development for Targeting Human Diseases

Editor: Israel Valencia Quiroz

ISSN (Online): 2468-5372

ISSN (Print): 2468-5364

ISBN (Online): 978-981-5223-16-3

ISBN (Print): 978-981-5223-17-0

ISBN (Paperback): 978-981-5223-18-7

First published in 2024.

need for a court order if at any point you breach any terms of this License Agreement. In no event will any delay or failure by Bentham Science Publishers in enforcing your compliance with this License Agreement constitute a waiver of any of its rights.

3. You acknowledge that you have read this License Agreement, and agree to be bound by its terms and conditions. To the extent that any other terms and conditions presented on any website of Bentham Science Publishers conflict with, or are inconsistent with, the terms and conditions set out in this License Agreement, you acknowledge that the terms and conditions set out in this License Agreement shall prevail.

Bentham Science Publishers Pte. Ltd.
80 Robinson Road #02-00
Singapore 068898
Singapore
Email: subscriptions@benthamscience.net

**BENTHAM
SCIENCE**

CONTENTS

PREFACE

Biotechnology weaves its transformational effect across the intricate fabric of life, altering our understanding of illnesses and igniting creative approaches to their treatment. Intricate biological processes and disease mechanisms have been decoded through the groundbreaking use of biotechnology methods. The potential of biotechnology in the context of drug development is considerable. "Recent advances in Biotechnology Vol. 9, Biotechnology and Drug Development for Targeting Human Diseases" is a collection of insightful discussions and in-depth analyses on the use of biotechnology in the treatment of disease.

The chapters in this book provide a thorough examination of the many biotechnology applications, covering topics like the use of multi-omics profiles in disease research and drug development, *in silico* drug design techniques, the use of viruses as carriers, and the investigation of natural products for use in wound healing and as antimicrobials. The notion of drug repurposing, the intersection of omics technologies with biotechnology in drug interaction investigations, and the most recent biotechnological discoveries in disease prevention also receive special emphasis.

Every chapter in this book has been meticulously chosen to provide thorough, up-to-date, and understandable knowledge, backed by a variety of references that let the reader dive deeper into the subject. These chapters work together to give readers a comprehensive picture of how biotechnology is fundamentally changing the field of drug research and disease treatment.

We sincerely thank the authors of each chapter for their contributions to the spirit of this book through their knowledge and thorough study. This compilation was made possible by their perceptions, know-how, and diligence.

We would like to express our sincere gratitude to our families, whose unwavering support has been essential at every stage of the process of writing this book. Their support and faith in this project have been essential.

We really hope that this book will be a helpful resource for individuals who are curious about the field of biotechnology and its applications to the treatment of diseases. This information not only sheds light on the condition of the field now, but also prepares the path for future developments in biotechnology and pharmaceutical research.

Israel Valencia Quiroz
Phytochemistry Laboratory, UBIPRO
Superior Studies Faculty (FES)-Iztacala
National Autonomous University of México (UNAM)
Tlalnepantla de Baz, México State, 54090
México

List of Contributors

Adriana Montserrat Espinosa-González — Phytochemistry Laboratory, UBIPRO, Superior Studies Faculty (FES)-Iztacala, National Autonomous University of México (UNAM), Tlalnepantla de Baz, México State, 54090, México

Axel R. Molina-Gallardo — Laboratory of Natural Products Bioactivity, UBIPRO, Superior Studies Faculty (FES)-Iztacala, National Autonomous University of México (UNAM), Tlalnepantla de Baz, México State, 54090, México

Ana K. Villagómez-Guzmán — Laboratory of Natural Products Bioactivity, UBIPRO, Superior Studies Faculty (FES)-Iztacala, National Autonomous University of Mexico (UNAM), Tlalnepantla de Baz, México State, 54090, México

Ana María García-Bores — Phytochemistry Laboratory, UBIPRO, Superior Studies Faculty (FES)-Iztacala, National Autonomous University of Mexico (UNAM), Tlalnepantla de Baz, Mexico State, 54090, Mexico

Carlos Pérez-Plasencia — Genomics Lab, National Cancer Institute (INCan), Tlalpan, Mexico City, 14080, Mexico
Genomics Lab, Biomedicine Unit, FES-Iztacala, National Autonomous University of Mexico, Tlalnepantla, 54090, Mexico

C. Tzasna Hernández-Delgado — Laboratory of Natural Products Bioactivity, UBIPRO, Superior Studies Faculty (FES)-Iztacala, National Autonomous University of México (UNAM), Tlalnepantla de Baz, México State, 54090, México

Erick Nolasco-Ontiveros — Phytochemistry Laboratory, UBIPRO, Superior Studies Faculty (FES)-Iztacala, National Autonomous University of México (UNAM), Tlalnepantla de Baz, México State, 54090, México

Eduardo López-Urrutia — Genomics Lab, Biomedicine Unit, FES-Iztacala, National Autonomous University of Mexico, Tlalnepantla 54090, Mexico

Edgar Antonio Estrella-Parra — Phytochemistry Laboratory, UBIPRO, Superior Studies Faculty (FES)-Iztacala, National Autonomous University of Mexico (UNAM), Tlalnepantla de Baz, México State, 54090, México

Felix Krengel — Department of Ecology and Natural Products, Faculty of Sciences, National Autonomous University of Mexico (UNAM), Coyoacan, Mexico City, 04510, Mexico

Ignacio Peñalosa Castro — Phytochemistry Laboratory, UBIPRO, Superior Studies Faculty (FES)-Iztacala, National Autonomous University of Mexico (UNAM), Tlalnepantla de Baz, México State, 54090, México

Israel Valencia Quiroz — Phytochemistry Laboratory, UBIPRO, Superior Studies Faculty (FES)-Iztacala, National Autonomous University of México (UNAM), Tlalnepantla de Baz, México State, 54090, México

Jose Cruz Rivera Cabrera — Liquid Chromatography Laboratory, Department of Pharmacology, Military School of Medicine, CDA, Palomas S/N, Lomas de San Isidro, 11200, México City, México

Julieta Orozco-Martínez — Laboratory of Natural Products Bioactivity, UBIPRO, Superior Studies Faculty (FES)-Iztacala, National Autonomous University of México (UNAM), Tlalnepantla de Baz, México State, 54090, México

José Guillermo Avila-Acevedo	Phytochemistry Laboratory, UBIPRO, Superior Studies Faculty (FES)-Iztacala, National Autonomous University of México (UNAM), Tlalnepantla de Baz, México State, 54090, México
José del Carmen Benítez-Flores	Histology Laboratory 1, UMF, Superior Studies Faculty (FES)-Iztacala, National Autonomous University of Mexico (UNAM), Tlalnepantla de Baz, Mexico City, Mexico State, 54090, México
Juan Carlos Gómez-Verjan	National Institute of Geriatrics (INGER), Blvd. Adolfo Ruiz, Cortines 2767, México City, 10200, México
Mai M. Badr	Department of Environmental Health, High Institute of Public Health (HIPH), Alexandria University, Alexandria, Egypt
María del Socorro Sánchez-Correa	Scientific Research Laboratory I, Superior Studies Faculty (FES)-Iztacala, National Autonomous University of México (UNAM), Tlalnepantla de Baz, México State, 54090, México
Nallely Álvarez-Santos	Phytochemistry Laboratory, UBIPRO, Superior Studies Faculty (FES)-Iztacala, National Autonomous University of Mexico (UNAM), Tlalnepantla de Baz, Mexico State, 54090, Mexico Postgraduate Biological Sciences, Postgraduate Studies Unit, National Autonomous University of Mexico (UNAM), Coyoacan, Mexico City, 04510, Mexico
Nadia Alejandra Rivero-Segura	National Institute of Geriatrics (INGER), Blvd. Adolfo Ruiz Cortines 2767, México City, 10200, México
Olivia Pérez-Valera	Institute of Chemistry, National Autonomous University of Mexico (UNAM), Mexico City, 04510, Mexico
Patricia Guevara-Fefer	Department of Ecology and Natural Products, Faculty of Sciences, National Autonomous University of Mexico (UNAM), Coyoacan, Mexico City, 04510, Mexico
Rafael Torres-Martínez	Chemical Ecology and Agroecology Laboratory, Research Institute for Ecosystems and Sustainability, National Autonomous University of Mexico (UNAM), Morelia, Michoacan, Mexico
Rocío Serrano-Parrales	Laboratory of Bioactivity of Natural Products, UBIPRO, Superior Studies Faculty (FES)-Iztacala, National Autonomous University of Mexico (UNAM), Tlalnepantla de Baz, México State, 54090, México
Verónica García-Castillo	Genomics Lab, Biomedicine Unit, FES-Iztacala, National Autonomous University of Mexico, Tlalnepantla 54090, Mexico
Viridiana R. Escartín-Alpizar	Laboratory of Natural Products Bioactivity, UBIPRO, Superior Studies Faculty (FES)-Iztacala, National Autonomous University of México (UNAM), Tlalnepantla de Baz, México State, 54090, México
Yesica R. Cruz-Martínez	Natural Products Bioactivity Laboratory, UBIPRO, Superior Studies Faculty (FES)-Iztacala, National Autonomous University of México (UNAM), Tlalnepantla de Baz, México State, 54090, México
Yuri Córdoba-Campo	Manuela Beltran University, Bucaramanga, Colombia

CHAPTER 1

Multi-omics Profiles are Applicable to Human Diseases and Drug Development

Adriana Montserrat Espinosa-González[1], José del Carmen Benítez-Flores[2], Juan Carlos Gómez-Verjan[3], Nadia Alejandra Rivero-Segura[3], Ignacio Peñalosa Castro[1], Jose Cruz Rivera Cabrera[4] and Edgar Antonio Estrella-Parra[1,*]

[1] *Phytochemistry Laboratory, UBIPRO, Superior Studies Faculty (FES)-Iztacala, National Autonomous University of Mexico (UNAM), Tlalnepantla de Baz, México State, 54090, México*

[2] *Histology Laboratory 1, UMF, Superior Studies Faculty (FES)-Iztacala, National Autonomous University of Mexico (UNAM), Tlalnepantla de Baz, Mexico City, Mexico State, 54090, México*

[3] *National Institute of Geriatrics (INGER), Blvd. Adolfo Ruiz Cortines 2767, México City, 10200, México*

[4] *Liquid Chromatography Laboratory, Department of Pharmacology, Military School of Medicine, CDA, Palomas S/N, Lomas de San Isidro, 11200, México City, México*

Abstract: Traditional medicine has been a reliable source for the discovery of molecules with therapeutic activity against human diseases of clinical interest. In the past, knowledge of traditional medicine was mainly transmitted orally and in writing. Recently, the advent of "multiomics" tools (transcriptomics, metabolomics, epigenomics, proteomics, and lipidomics, among others) has increased and merged our knowledge, both traditional knowledge and that gained with these new multiomics technologies. In this way, the development of medicines with these 'multiomics technologies' has allowed pharmaceutical advances in the discovery of new drugs. In addition, 'multiomics' technologies have made it possible to uncover new biological activities of drugs that are currently used in clinical therapy. In the same way, 'multiomics' has allowed for the development of 'personalized medicine', that is, a particular and specific treatment and/or diagnosis of a patient with respect to a disease. Therefore, 'multiomics' technologies have facilitated the discovery of new clinical therapeutics for disease, as well as allowing for the diagnosis and/or treatment of diseases in an individual and personalized way.

* **Corresponding author Edgar Antonio Estrella-Parra:** Phytochemistry Laboratory, UBIPRO, Superior Studies Faculty (FES)-Iztacala, National Autonomous University of Mexico (UNAM), Tlalnepantla de Baz, México State, 54090, México; Tel: +525556231136; E-mail: estreparr@iztacala.unam.mx

Israel Valencia Quiroz (Ed.)

Keywords: Drug development, Multiomics technology, Medicinal traditional, Personalized medicine.

INTRODUCTION

In the past, knowledge from original peoples was transmitted only from generation to generation, but today, the knowledge is used in the development of new medicines [1]. Natural products are chemical compounds produced by living organisms, including plants, animals, and microorganisms, and have long been used in medicine and other biological applications [1]. Biological research has undergone many changes since the end of the 20th century and the beginning of the 21st century, with the publication of the complete human genome sequence by the International Genome Sequencing Consortium in 2003 being a crucial step in genetic research [2]. In a similar manner, drug development has been considered a conservative strategy with highly regulated processes. However, medicine is rapidly evolving with the help of different strategies that allow for the development of comprehensive and personalized treatments for different types of diseases and/or patients [3].

The omic sciences are a set of technologies used to study the global molecular components of an organism, such as genes, proteins, metabolites, and lipids. These technologies include genomics, transcriptomics, proteomics, metabolomics, and lipidomics; furthermore, these technologies have been used in a wide variety of applications, including research in biology, medicine, agriculture, and ecology [3, 4]. These "omics technologies" and advances in bioinformatics have generated new knowledge and integrated new technologies such as artificial intelligence (AI) to improve precision medicine [5]. In this "post genomics" era, research is focused on the role of genes, understanding transcriptional regulation, the biochemical roles of gene products and their interactions, and understanding how various chemicals influence metabolic behavior. These new "omics" technologies are based on global and high-throughput analytical methods, such as microarrays, 2D-gel, 2DLC/MS and mass spectrometry, which produce data on a large scale, as well as bioinformatics and computer modeling [2, 3]. In this manner, multiomics sciences are used to identify and investigate new bioactive compounds from natural products [3, 6].

Important factors for the success of precision medicine (or personalized medicine) include early clinical development, the "back translation" of knowledge *via* the development of drugs and the translation of omic signatures into clinically relevant biomarkers, as well as the development of precision diagnostics adapted to each patient [3]. Moreover, multiomics science permits the development of these omic technologies and their application in biomedical research and pharmaceutical products, thereby offering a broader exploration of the genome, transcriptome, and proteome and with a greater possibility of finding solutions for the discovery and validation of new drugs, evaluating their efficacy, toxicity, safety and personalized access, as well as the availability of new drugs [2].

The goal of this chapter is to describe the development of new drugs used in clinical therapy and their applicability in personalized medicine based on multiomics sciences.

ANCESTRAL KNOWLEDGE: TRADITIONAL MEDICINE IN THE MULTIOMICS ERA

There is a growing interest in the discovery of new drugs from traditional medicine [7]. Ancestrally, knowledge has been transferred from generation to generation, although in modern times, this knowledge that is transferred orally is at risk of being lost [1], not only hindering the development of new drugs but also the discovery of new therapeutic strategies [8]. Ancestral documents such as the 'Shenlong's classis of materia medica' from China describe the use of 365 drugs; moreover, in ancient Greece, Dioscorides described the use of 600 medicinal plants with therapeutic activity [9]. In medieval Europe, traditional medicine comes from the Greeks and Romans such as Hippocrates, Galen, and Dioscorides, and this knowledge was preserved by Benedictine monks through botanical gardens such as the Abbeys of Montecassino and St. Gall, respectively [10]. A convergent referent between traditional medicine and omics science occurred in Japan. In this country, Chinese medical practice was introduced in the 6th century A.D., and eventually the concept of 'KAMPOmics', which represented the merging of omic sciences with traditional Japanese medicine, was developed [11]. The principles of yin (cold) and yang (hot) in traditional Chinese medicine were evaluated using metabolomics on serum from fever rats administered a traditional herbal treatment. The rats had an increase in temperature following treatment with plants that stimulated heat, in contrast to their response following treatment with plants that reduce temperature; certain metabolomic markers could discriminate the samples based on the traditional herbal treatment [12]. In addition, in 2014, the Brazilian government published a book that summarized the traditional medicine of the '*Yanomani people*', which identifies the botanical species and their preparations that are used as therapeutic material [8]. Furthermore, the 'Herbalomic project', which focuses on new methods to elucidate molecules, establishes libraries of plants in the context of traditional Chinese medicine [13]. Concurrently, China developed the concept of GP-TCM (Good Practice in Traditional Chinese Medicine research in the postgenomic era), which utilizes coordinated actions to regulate interdisciplinary and intersectoral activities in traditional medicine [14].

Consequently, in the 20th century and early 21st century, innovations have been made that help us to understand life [15]. Traditional natural medicines can be modernized with the use of novel high-tech methods for the development of new phytotherapeutics [1]. In this way, the FDA of the USA describes omics science as a technological tool with automated methods to analyze several types of molecules simultaneously [16], with a methodological strategy for the study, standardization and quality control of herbal formulas [17]. Accordingly, omics technology, such as genomics, transcriptomics, proteomics and metabolomics, helps us understand the pharmacologic effects of plants used in traditional medicine [8].

Consequently, there is an important relationship between ancestral knowledge in medicine and omics tools, and this relationship has led to work that brings together traditions and innovative technologies.

GLOBALIZATION OF TRADITIONAL MEDICINE

Previously, due to the effects of globalization, plants were only used locally, and their use outside the local population was restricted. However, recently, the globalization of traditional medicines has led to self-medication in which herbal remedies such as '*aryuveda*' and other therapies appear in supermarkets, health stores, and pharmacies, among other places of business [18].

Currently, the economic interest in herbal remedies as alternative and complementary medicines in the United States is estimated to be approximately 50-128.8 million dollars [16]. Moreover, the globalization of herbal medicine products affects the market within the USA, and there must be communication between the scientific community and industry [9]. Therefore, the pharmaceutical industry cannot ignore emerging markets in the development of new therapeutic substances because this information can be used to reduce costs and the number of obstacles preventing their approval [19]. Pharmaceutical and biotech companies often confidentially apply translational emerging safety biomarkers (ESBs) during drug development, which influences the development of new drugs [20].

Thus, the search for new therapies for various diseases has given rise to a greater diffusion of traditional medicine, even outside the place of origin, particularly in the search for molecules with therapeutic activity, as we will see later.

THE CONSTRUCTION OF NEW DRUGS BASED ON THE OMICS APPROACH

The postgenomic era started with the completion of the Human Genome Project [13], and new drugs are continually being developed [21]. There have been success stories in the development of new drugs; for example, antiretroviral therapy against HIV/AIDS decreased mortality from 16.2 (1995) to 2.7% (2010), and medicines related to heart disease reduced mortality by 45% from 1999 to 2005 [19]. Moreover, between 1981 and 2014, many new drugs were introduced into the market, more than 50% of which came from natural products [8]. In this manner, in 2015, a researcher named you-you was awarded the Nobel prize for the discovery of 'artemisinin', a drug to fight malaria. This compound was extracted from *Artemisia annua* L., which is a plant used in traditional Chinese medicine [22] that was reproduced based on a recipe from an ancient prescription handbook [16]. Other drugs used in clinical therapy, such as captopril, enalapril and lisinopril, were developed based on peptides that were isolated from the Brazilian snake *Bothrops jararaca*, as well as the anti-malaria drug malarone, which was a model Lapachol molecule isolated from a tree that was used in traditional Brazilian medicine [8]. Other drugs, such as pirfenidone and nintedanib, are antifibrotic agents that increase the risk of idiopathic pulmonary fibrosis but have adverse effects during treatment; the natural product galectin-3 is a promising agent with beneficial effects and is currently undergoing phase 2 clinical trials [23].

Novel technologies make the development of a new drug more efficient, but they also lead to more detailed requirements, which increase the time and economic costs required to implement the latest generation of drugs [19]. Recently, the use of omics techniques for scientific research has increased, and omics can be used to analyze most classes of biological molecules, such as DNA, RNA, proteins and metabolites [24]. Omics techniques are useful for the identification of biological targets and the elucidation of mechanisms of action in drug discovery [25]. For example, omics tools have allowed us to identify the differences in breast cancer in two female patients; the results showed that there were differences between the two individuals at multiple biological levels [26]. Omics tools can also be used to evaluate comorbidities and differences in various types of gastrointestinal tract cancers [27]. Moreover, omics tools were used to determine that histone H1 regulates chromatin compaction in humans, as well as the mechanisms of transcription and coregulation [28]. Analyses using omics have allowed us to establish the concept of 'deep phenotyping', which refers to defining the biological age and classifying the human body by groups of organs and systems, with the goal of inspecting the longevity of people [29]. Along this line, the InnoMed PredTox consortium (PredTox Project) was created to ensure safety in preclinical studies by incorporating multiomics tools from real-life data, as well as data about drug candidates from various participating companies that previously failed during nonclinical development [30].

More than half of all diagnosed lung cancer patients are in a very advanced stage or in metastasis; thus, it is necessary to determine biomarkers in the initial stage using multi-omics tools such as genomics, transcriptomics, and metabolomics, which would discriminate between malignant and benign nodules or simple injuries [31]. A certain diagnosis is necessary for a good prognosis and quality of life, as we will see later.

OMICS SCIENCE IN PERSONALIZED MEDICINE: A GOLD STANDARD

'Personalized medicine' is a recent concept that investigates how differences between individuals affect the way they respond to a drug. Personalized medicine is a strategy that prevents incorrect diagnoses and applies optimal treatments for a particular disease [32], thereby providing a better patient prognosis [33]. Moreover, personalized medicine accounts for variability in the molecular, genomic, cellular, clinical, environmental and physiological dimensions [34, 35]. Tools such as multiomics and bioinformatics provide an opportunity for a good prognosis for patients with various diseases [36] and are considered the *'gold standard'* [37]. These types of omics tools allowed for the development of 'biomarkers', which are molecules that are measured before and after exposure to a medical product and are important for the development of new drugs [37, 20].

The contributions of omics to the development of personalized medicine have been widely reported. In the treatment of asthma and COPD, pharmacological and ventilator treatments have not changed over five decades; meanwhile, personalized medicine not only includes traditional treatment but also treats symptoms and aids in the development of drugs for these conditions [38]. In patients with asthma (from medium to severe), transcriptomic and proteomic analysis was carried out on 266 people, and their profiles were compared with that of the omics database; these data were used to define a phenotype that is associated with smokers, and these essential tools allow for a more personalized treatment according to 'your omics profile' [39]. Cholangiocarcinoma is a rare cancer; based on omics data and biomarkers discovered from transcriptomic studies, research on the identification of candidate drugs to treat this type of rare cancer accelerated [40]. Meanwhile, the multiomics profile of 155 esophageal squamous cell carcinoma samples allowed for the accurate diagnosis of cancer patients and the prediction of therapeutic response with 85.75% sensitivity and 90% specificity; these data allowed us to distinguish the four subtypes of dominant alterations and predicting a possible personal therapy [41]. The construction of an inclusive multiomics model was used to monitor breast cancer based on the clinical data from several patients; this data was used to provide more feasible diagnoses [42]. In neurodegenerative diseases, lipid variability was observed in the plasmalemmas, as well as deregulation of lipid metabolism, particularly in the growth cones, such as the lipids lysophosphatidylserines and cardiolipins, which could be possible biomarkers in neurodegenerative diseases [43]. In addition, omics tools even made it possible to identify 83 genes that are associated with both PD and breast cancer, which allowed for the more efficient prediction of specific drugs that would be effective for these diseases [44]. In necrotizing enterocolitis, multiomics analysis allowed for the discovery of biomarkers of this disease; the biomarkers were identified using available information and processed by algorithms [33]. In Parkinson's disease (PD), multiomics tools helped develop therapeutic strategies for this condition through epigenetic analysis, as well as personalized nutrition, which contributes to this disease [45]. Furthermore, in the early diagnosis of PD, biomarkers such as α-synuclein combined with enhanced T2 star weighted angiography and microRNA-4639-5p were identified from proteomic profiles [46]. In papillary thyroid cancer, six proteins (FYN, JUN, LYN, PML, SIN3A, and RARA) and the Erb-B2, CDK1 and CDK2 receptors, as well as histone deacetylase receptors, were identified using multiomics tools; these proteins and other miRNAs were found to be biomarkers for this disease [47]. Two subtypes of lactate metabolism patterns were established in lung squamous carcinoma, and the application of a prognostic index (LMRPI) predicted the prognosis of the disease based on synergy with some anticancer therapies [48]. In addition, 1,061 biomarkers and 892 constitutive biomarkers were identified in the plasma of patients in the acute posttraumatic phase [49]. A multiomics deep learning network method was used to distinguish glioma patients with poor prognoses that are in the dire need of treatment through the construction of transcriptome, miRNA, and DNA methylation profiles, among others; in this case, omics tools helped find drug targets for different gliomas [50]. In patients with glioma, inhibitory CDH11 methylation was found to contribute to poor prognosis [50]. In patients with liver metastasis, information about the immune microenvironment of cancer cells was determined; the cells had high levels of T-cell suppression and other markers, which were useful for predicting a good prognosis [51]. In 80% of acute lymphoblastic leukemia patients, oncogenic lesions were identified, as well as nonconductive mutations at the subclonal level; this allowed researchers to infer resistance to cancer therapy, and in the future, this information could be used to establish personalized therapy for this disease [52]. Furthermore, the bone marrow microenvironment of acute myeloid leukemia patients was analyzed using the secretome/transcriptome, and the identification of deregulated genes (Tfpi, Dtk, KLKB1, and Prekallikrein) and proteins led to the conclusion that the microenvironment is active in this disease [53]. However, importantly, bioethics in studies with omics research must adhere to human rights and the principles of every person, justice, and charity [35].

Medicines have never been more personalized than they are now. Coupled with technological development, personalized medicine allows for a better patient prognosis, but this has also necessitated the search for new drugs as an omics approach.

CHALLENGES IN THE DISCOVERY OF NEW DRUGS

The development of drugs and techniques in some areas of health has lagged behind for years [38]. Thus, novel technologies make the development of new drugs more efficient, but it has led to more detailed requirements and an increase in the time and economic cost required to implement the latest generation of drugs [19]. The pharmaceutical industry and the scientific community have worked jointly through the use of omic tools to develop and discover new drugs with lower cost and time requirements for their application [16]. In this manner, omics tools such as genomics, proteomics and metabolomics can lead to the discovery of active molecules [54]. Additionally, multiomics tools such as gene-centric multichannel (GCMC) have been used to predict cancer drug response, and these models determined the efficacy of 265 drugs used for cancer therapy [55]. Additionally, genome-wide association studies (GWASs) allow for the identification of variants and associated loci in various diseases, thereby providing information for drug development [42]. In contrast, the use of omics tools to understand the mechanisms of idiosyncratic drug-induced hepatotoxicity demonstrated that idiosyncratic drugs induce an increase in intercellular ceramides, which changes the expression of genes by inducing inflammation and ER stress [56].

Recently, there has been much interest in studying drugs already established as therapy for other diseases. Currently, the pharmaceutical industry is looking for molecules that interact simultaneously and specifically with multiple therapeutic targets, a term called 'compound promiscuity' [57]. Therefore, 'empagliflozin', which is used in diabetes and patients with obesity, was explored using omics tools; the results showed that this drug modulates the microbiota and the metabolism of tryptophan, making it a promising drug against obesity based on the host-microbe interaction [58]. 'Capreomycin' is a drug used to treat tuberculosis; multiomics analysis showed mutations in tlYA in some drug-resistant strains, as well as dysregulation of lipid and fatty acid metabolism. This result will allow for the readjustment of therapeutic treatments for tuberculosis [59]. Another drug, triclosan, was evaluated by metabolomic analysis and the results showed that it induces hepatoxicity and enterotoxicity [60]. Likewise, cyclosporin-A induces cholestiasis [21] and mitochondrial damage by activating Nrf2 and ATF4 [61]. Likewise, in breast cancer, growth differentiation factor 10 (GDF10) is associated with the progression of breast cancer and is a promising target for the development of drugs [62]. Additionally, through the proteomic analysis of at least 949 cancer cell lines from 28 different types of tissue, the synergy of various drugs in cancer therapy was analyzed. As there were only 1500 proteins with potential predictive power for this disease, a proteomic pan cancer map was developed [63]. In invasive breast carcinoma (BRCA), multiomics approaches have been used to identify potential autophagy regulators, such as SF3B3, TRAPPC10, SIRT3, MTERFD1, and FBXO5, with SF3B3 and SIRT3 being new targets for drug development [6]. In glioblastoma, the FN1 biomarker was discovered and found to have many implications in this disease; the FN1 molecule is a marker of a good prognosis in the initial stages of this disease [64]. The mechanism of action of the new antimalarial compound JPC-3210 (2-aminomethylphenol), which is in the final stages of preclinical development prior to testing in humans through proteomic, metabolomic and peptidomic analysis, was elucidated, and the mechanism included the inhibition of hemoglobin and the deregulation of DNA replication and the translation of *Plasmodium falciparum* proteins [65]. In atherosclerosis, the interactome between the intestinal microbiota and antibiotics induces a loss of intestinal diversity, decreases tryptophan abundance, and alters lipid metabolism [66]. Additionally, omics technologies have allowed for advances in studies on the treatment of osteoarthritis, which has permitted the development of new drugs for the disease [67].

Furthermore, the study of natural products for the development of new pharmaceuticals is continuous. Thus, the use of traditional Chinese medicine with multiomics tools has allowed for the identification of biomarkers such as ERBB2, MYC, FLT4, TEK, GLI1, TOP2A, PDE10A, SLC6A3, GPR55, TERT, EGFR, KCNA3 and HDAC4, which are differentially expressed in different human cancer cell lines [68]. Additionally, the natural compound luteolin-7-O-a-L-rhamnoside is a potential 'promiscuous enzyme inhibitor' of tyrosinase, hyaluronidase and alpha amylase, and is implicated in some chronic diseases [57]. In addition, metabolomic and proteomic analyses allowed us to determine the profile of molecules in ischemic stroke and their interaction with a decoction of Chinese medicinal plants; in this study, researchers determined the neuroprotective effects of molecules such as scutellarin, quercetin 3-O-glucuronide, ginsenoside Rb1, schizandrol A and 3,5-diCQA, which activate the NF-kB signaling pathway [69]. The decoction used in traditional Chinese medicine was from the 'Qing dynasty', and it improved brain function in a model of cerebral ischemia. Using proteomics, metabolomics and transcriptome methods, 15 targets, such as Aprt, Pde1b, Gpd1, Glb1, HEXA and HEXB, were found to reverse the adverse effects of cerebral ischemia [70]. In chronic obstructive pulmonary disease, a traditional Chinese medicine decoction (bufei Jianoi granules) reduced the duration of acute exacerbation of the disease; proteomics, metabolomics and bioinformatics analyses showed that natural products such as pachymic acid, shionone, peiminine and astragaloside A activated the EGFR, ERK1, PAI-1, and p53 signaling pathways, making them promising agents for the development of new drugs [71]. Through transcriptomic profiling, the alkaloid roemerin was found to have activity against *Bacillus subtilis*; roemerin accumulated in cells and generated oxidative stress and ROS [25]. Moreover, integrative omics studies identified 29 compounds from plants, such as luteolin, apigenin, and thujone, among others, that were bioactive against non-small cell lung cancer and aided in the discovery of differences in disease types and the prediction of potential therapeutic strategies [72].

Hence, the information obtained by clinical analyses with recent technology has led to the discovery of new molecules with therapeutic activity against particular diseases.

BIOINFORMATICS IN OMICS: AN ACCUMULATION OF EXPERIENCES IN SYNERGY

None of the recent medical advances would be contextualized if there was no technological support and without the development of bioinformatics. The current technologies have allowed for the development of phage-nomic, epigenetic, proteomic, and metabolomic data, although assembling such information is a challenge [73]. Currently, there is an effort to combine omics data with clinical data to create several databases and computer programs [34]. 'IntelliOmics' is a term that allows for the complete analysis of raw data files until a diagnostic report is obtained, which can be associated with the treatment recommendation [74]. 'Automics' allows for the integration of omics tools into algorithmic models through the construction of unique omics models for each data, which are then combined in a deep learning mathematical program [73].

In 2014, the National Cancer Institute allowed access to the Cancer Genome Atlas (TCGA) database, which was created using various omics tools derived from the analysis of cancer patients [26]. Through bioinformatics analysis, which used data from the Cancer Genome Atlas Database and Molecular Signatures Database (508 patients), transcriptomics and genomics helped to better predict the prognosis of invasive ductal carcinoma of the breast [75]. The cancer therapeutics database response portal allowed for the accurate prediction of drug response in different tissues with cancer and correlations between their genomic and molecular characteristics in response to various drugs were assessed [76]. The myocardial infarction knowledge base (MIKB) is a database that includes 1,782 omics factors, 28 MI subtypes, and 2,347 omics factor-MI interactions, as well as 1,253 genes and 6 chromosomal alterations collected from 2,647 research articles [77]. Additionally, for the discovery of drugs, GWAS meta-analysis has been used in combination with genomic studies, as in the case of thromboembolism, in which the target molecules are interleukin-4 and interleukin-13 [78]. The Omics and Multidimensional Spatial (OMS) method is a method that has been used to evaluate the clinical metadata of different patients who present with therapeutically resistant metastasis; this data has allowed for the identification of new therapeutic vulnerabilities that can lead to more effective treatments for cancer [79]. 'SUMO' is a computational program that refines factorization and multi-patient similarities to identify similar molecular subtypes of patients with any disease by reducing noise and improving incomplete data [80]. Thus, in patients diagnosed with lower-grade glioma, 'SUMO' was used to determined that non-CpG island methylation is associated with the gene CLCF1, and which is a biomarker of glioma [80]. In HIV-1 infection, using 'SWATH-MS' analyses and proteomic data, three factors (LAG-3, CD147, CD231) were found to be altered in several infected cell lines, thereby confirming that there is a universal antigen because of the variability in biomarkers between the different clones [81]. The samples of 4,277 healthy subjects were collected to discriminate the basal levels of optimal health using algorithms [36]. Moreover, docking analyses revealed that 5 drugs could be helpful in the treatment of papillary thyroid cancer disease [47].

In neurodegenerative diseases, samples of patients with amyotrophic lateral sclerosis were analyzed using omics databases and computational methods, and different phenotypes and deregulated pathways of the disease were discovered [82]. 'CPAS' is a series of algorithms that allow for multiomics analysis of copy number variation (CNV) genes, allowing for the identification of biological pathways that would be undetectable by simple omics analysis [83]. The SIT-DIMS analysis platform, through the use of algorithms and drug libraries of cancer patients, can establish phenotypic databases and quantify synergy for the discovery of combinatorial strategies [84]. With the multiomics integration program 'MASPD' and proteomic data, the proteins present in the microcellular domains were identified, as well as their gene expression, in patients with schizophrenia [85]. 'OmicView' is a visualization platform that allows for the identification of biomarkers that interact with any drug [86]. 'MOGSA' is another computational method for the analysis of data from only a single omics method, and this program integrates multiple experimental data types [87]. 'iProFun' is a multiomics computer program that allows for the analysis of altered and methylated DNA in tumor samples. In very aggressive ovarian cancer, 600 genes with methylation and copy number alterations were identified, and the AKT1 oncogene was found to interact as a node in the cancer process; thus, this type of analysis provides biological information for the development of drugs [88]. Through machine learning and omics tools, biomarkers of clinical interest in cell carcinoma, such as ATP4B, AC144831.1 and Tfcp211, were identified [89]. ClusterProfiler 4.0 is a bioconductor package that supports omics data from thousands of organisms based on internal ontologies and pathways derived from online databases [90]. 'Panomicon' is a web based platform that performs multiomics analysis, and improves the storage and management of omic data, as well as their visualization and the interactions of the different omic tools [91]. Likewise, through the use of libraries such as UALKAN, KM plotter, and others, the analysis of invasive breast carcinoma tissues showed that the TP53 gene is mutated in 30% of samples, with overexpression of this gene slowing pharmacological effectiveness, making it a potential biomarker for the development of drugs for anticancer therapy [92]. The databases 'R-ODAF' [93] and 'TRANSFAC' [94] are used to process old data by taking data from old microarrays and from other state-of-the-art platforms to increase the certainty of transcriptomic analysis results.

Therefore, the development of new drugs and personalized medicine could not be conceived of without the development of bioinformatics, which acts in synergy *in vivo* and *in silico*.

In summary, the synergy between ancestral knowledge, omics science, the development of new drugs, bioinformatics/databases, personalized medicine, and the original people of this ancestral knowledge of traditional medicine (Fig. **1**) is essential to understanding the different areas of knowledge in the search for promising drugs for clinical therapy.

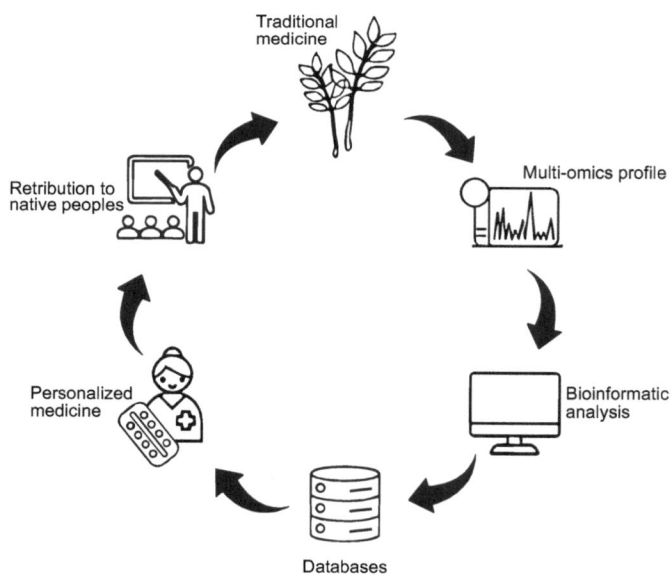

Fig. (1). Interaction and synergy between traditional medicine, omics sciences, bioinformatics/databases, and personalized medicine in the search for new drugs.

CONCLUSION

Many strategies can be used to achieve only one objective. Ancient knowledge in synergy with new technologies, particularly based on traditional medicines, omics tools, bioinformatics/machine learning, and the subsequent development of drugs with therapeutic activity, together can be used for the successful development of medicines, especially in the era of developing new drugs *via* a multiomics approach.

REFERENCES

[1]   Efferth T. Perspectives for Globalized Natural Medicines. CJNM 2011; 9(1): 1-6.

[2]   Yan SK, Liu RH, Jin HZ, *et al.* "Omics" in pharmaceutical research: overview, applications, challenges, and future perspectives. Chin J Nat Med 2015; 13(1): 3-21.
[http://dx.doi.org/10.1016/S1875-5364(15)60002-4] [PMID: 25660284]

[3]   Hartl D, de Luca V, Kostikova A, *et al.* Translational precision medicine: An industry perspective. J Transl Med 2021; 19(1): 245.
[http://dx.doi.org/10.1186/s12967-021-02910-6] [PMID: 34090480]

[4]   Zhu X, Yao Q, Yang P, *et al.* Multi-omics approaches for in-depth understanding of therapeutic mechanism for Traditional Chinese Medicine. Front Pharmacol 2022; 13: 1031051.
[http://dx.doi.org/10.3389/fphar.2022.1031051] [PMID: 36506559]

[5]   Boniolo F, Dorigatti E, Ohnmacht AJ, Saur D, Schubert B, Menden MP. Artificial intelligence in early drug discovery enabling precision medicine. Expert Opin Drug Discov 2021; 16(9): 991-1007.
[http://dx.doi.org/10.1080/17460441.2021.1918096] [PMID: 34075855]

[6]   Zhang S, Zhang J, An Y, *et al.* Multi-omics approaches identify *SF3B3* and *SIRT3* as candidate autophagic regulators and druggable targets in invasive breast carcinoma. Acta Pharm Sin B 2021; 11(5): 1227-45.
[http://dx.doi.org/10.1016/j.apsb.2020.12.013] [PMID: 34094830]

[7]   Li S, Pei W, Yuan W, Yu D, Song H, Zhang H. Multi-omics joint analysis reveals the mechanism of action of the traditional Chinese medicine *Marsdenia tenacissima* (Roxb.) Moon in the treatment of hepatocellular carcinoma. J Ethnopharmacol 2022; 293(15): 115285.
[http://dx.doi.org/10.1016/j.jep.2022.115285] [PMID: 35429621]

[8]   Castro B. Brazilian traditional medicine: Historical basis, features and potentialities for pharmaceutical development. J Tradit Chin Med Sci 2020; 8(1): 44-50.

[9]   Werner K, Jacqueline W. The globalization of traditional medicines: Perspectives related to the european union regulatory environment. Eng J (NY) 2019; 5(1): 22-31.

[10]   Leonti M, Verpoorte R. Traditional mediterranean and european herbal medicines. J Ethnopharmacol 2017; 199(199): 161-7.
[http://dx.doi.org/10.1016/j.jep.2017.01.052] [PMID: 28179113]

[11]   Yamamoto M. KAMPOmics: A framework for multidisciplinary and comprehensive research on Japanese traditional medicine. Gene 2022; 831: 146555.
[http://dx.doi.org/10.1016/j.gene.2022.146555] [PMID: 35569769]

[12]   Leung Kwan KK, Wong TY, Wu QY, Xia Dong TT, Lam H, Keung Tsim KW. Mass spectrometry-based multi-omics analysis reveals the thermogenetic regulation of herbal medicine in rat model of yeast-induced fever. J Ethnopharmacol 2021; 279(279): 114382.
[http://dx.doi.org/10.1016/j.jep.2021.114382] [PMID: 34197959]

[13]   Haiyan L, Zhen G, York P, Shao Q, Jingkai G, Jiwen Z. Research of traditional chinese medicine in

terms of herbalomics. Wood Sci Technol 2010; 12(2): 160-4.
[http://dx.doi.org/10.1016/S1876-3553(11)60008-2]

[14]   Uzuner H, Bauer R, Fan TP, *et al.* Traditional Chinese medicine research in the post-genomic era: Good practice, priorities, challenges and opportunities. J Ethnopharmacol 2012; 140(3): 458-68.
[http://dx.doi.org/10.1016/j.jep.2012.02.028] [PMID: 22387462]

[15]   Mirowski P. The scientific dimensions of social knowledge and their distant echoes in 20th-century American philosophy of science. Stud Hist Philos Sci (35): 283-326.
[http://dx.doi.org/http://10.1016/j.shpsa.2003.11.002]

[16]   Pelkonen O, Pasanen M, Lindon JC, *et al.* Omics and its potential impact on R&D and regulation of complex herbal products. J Ethnopharmacol 2012; 140(3): 587-93.
[http://dx.doi.org/10.1016/j.jep.2012.01.035] [PMID: 22313626]

[17]   Buriani A, Garcia-Bermejo ML, Bosisio E, *et al.* Omic techniques in systems biology approaches to traditional Chinese medicine research: Present and future. J Ethnopharmacol 2012; 140(3): 535-44.
[http://dx.doi.org/10.1016/j.jep.2012.01.055] [PMID: 22342380]

[18]   Cordell GA. Phytochemistry and traditional medicine: The revolution continues. Phytochem Lett 2014; 10: xxviii-l.
[http://dx.doi.org/10.1016/j.phytol.2014.06.002]

[19]   Quan H, Chen X, Zhang J, Zhao PL. New paradigm for drug developments—From emerging market statistical perspective. Contemp Clin Trials 2013; 36(2): 697-703.
[http://dx.doi.org/10.1016/j.cct.2013.06.009] [PMID: 23810938]

[20]   Zabka TS, Burkhardt J, Reagan WJ, *et al.* The use of emerging safety biomarkers in nonclinical and clinical safety assessment: The current and future state: An IQ DruSafe industry survey. Regul Toxicol Pharmacol 2021; 120: 104857.
[http://dx.doi.org/10.1016/j.yrtph.2020.104857] [PMID: 33387566]

[21]   Van den Hof WF, Ruiz-Aracama A, Van Summeren A, *et al.* Integrating multiple omics to unravel mechanisms of Cyclosporin A induced hepatotoxicity *in vitro*. Toxycol in vitro 2015; 29(3): 489-501.

[22]   Gao R, Hu Y, Dan Y, Hao L, Liu X, Song J. Chinese herbal medicine resources: Where we stand. Chin Herb Med 2020; 12(1): 3-13.
[http://dx.doi.org/10.1016/j.chmed.2019.08.004] [PMID: 36117557]

[23]   Ghumman M, Dhamecha D, Gonsalves A, *et al.* Emerging drug delivery strategies for idiopathic pulmonary fibrosis treatment. Eur J Pharm Biopharm 2021; 164: 1-12.
[http://dx.doi.org/10.1016/j.ejpb.2021.03.017] [PMID: 33882301]

[24]   Verheijen M, Tong W, Shi L, Gant TW, Seligman B, Caiment F. Towards the development of an omics data analysis framework. Regul Toxicol Pharmacol 2020; 112: 104621.
[http://dx.doi.org/10.1016/j.yrtph.2020.104621] [PMID: 32087354]

[25]   Avci FG, Sayar NA, Sariyar Akbulut B. An OMIC approach to elaborate the antibacterial mechanisms of different alkaloids. Phytochemistry 2018; 149: 123-31.
[http://dx.doi.org/10.1016/j.phytochem.2017.12.023] [PMID: 29494814]

[26]   Aguilar B, Abdilleh K, Acquaah-Mensah GK. Multi-omics inference of differential breast cancer-related transcriptional regulatory network gene hubs between young Black and White patients. Cancer Genet 2023; 270-271: 1-11.
[http://dx.doi.org/10.1016/j.cancergen.2022.11.001] [PMID: 36410105]

[27]   Jiangzhou H, Zhang H, Sun R, *et al.* Integrative omics analysis reveals effective stratification and potential prognosis markers of pan-gastrointestinal cancers. iScience 2021; 24(8): 102824.
[http://dx.doi.org/10.1016/j.isci.2021.102824] [PMID: 34381964]

[28]   Ponte I, Andrés M, Jordan A, Roque A. Towards understanding the regulation of histone H1 somatic subtypes with OMICs. J Mol Biol 2021; 433(2): 166734.
[http://dx.doi.org/10.1016/j.jmb.2020.166734] [PMID: 33279581]

[29] Nie C, Li Y, Li R, *et al.* Distinct biological ages of organs and systems identified from a multi-omics study. Cell Rep 2022; 38(10): 110459.
[http://dx.doi.org/10.1016/j.celrep.2022.110459] [PMID: 35263580]

[30] Ellinger-Ziegelbauer H, Adler M, Amberg A, *et al.* The enhanced value of combining conventional and "omics" analyses in early assessment of drug-induced hepatobiliary injury. Toxicol Appl Pharmacol 2011; 252(2): 97-111.
[http://dx.doi.org/10.1016/j.taap.2010.09.022] [PMID: 20888850]

[31] Robles AI, Harris CC. Integration of multiple "OMIC" biomarkers: A precision medicine strategy for lung cancer. Lung Cancer 2017; 107: 50-8.
[http://dx.doi.org/10.1016/j.lungcan.2016.06.003] [PMID: 27344275]

[32] Gervasini G. Pharmacogenetics and personalized medicine. Are expectations being met? Med Clin 2019; 152(9): 368-71.
[http://dx.doi.org/10.1016/j.medcli.2018.12.001] [PMID: 30611536]

[33] Leiva T, Lueschow S, Burge K, Devette C, McElroy S, Chaaban H. Biomarkers of necrotizing enterocolitis in the era of machine learning and omics. Semin Perinatol 2023; 47(1): 151693.
[http://dx.doi.org/10.1016/j.semperi.2022.151693] [PMID: 36604292]

[34] Herr TM, Bielinski SJ, Bottinger E, *et al.* A conceptual model for translating omic data into clinical action. J Pathol Inform 2015; 6(1): 46.
[http://dx.doi.org/10.4103/2153-3539.163985] [PMID: 26430534]

[35] Williams JK, Anderson CM. Omics research ethics considerations. Nurs Outlook 2018; 66(4): 386-93.
[http://dx.doi.org/10.1016/j.outlook.2018.05.003] [PMID: 30001880]

[36] Zhang W, Wan Z, Li X, *et al.* A population-based study of precision health assessments using multi-omics network-derived biological functional modules. Cell Rep Med 2022; 3(12): 100847.
[http://dx.doi.org/10.1016/j.xcrm.2022.100847] [PMID: 36493776]

[37] Kulkarni S, Kannan M, Atreya CD. Omic approaches to quality biomarkers for stored platelets: Are we there yet? Transfus Med Rev 2010; 24(3): 211-7.
[http://dx.doi.org/10.1016/j.tmrv.2010.03.003] [PMID: 20656188]

[38] Chapman DG, King GG, Robinson PD, Farah CS, Thamrin C. The need for physiological phenotyping to develop new drugs for airways disease. Pharmacol Res 2020; 159: 105029.
[http://dx.doi.org/10.1016/j.phrs.2020.105029] [PMID: 32565310]

[39] Lefaudeux D, De Meulder B, Loza MJ, *et al.* U-BIOPRED clinical adult asthma clusters linked to a subset of sputum omics. J Allergy Clin Immunol 2017; 139(6): 1797-807.
[http://dx.doi.org/10.1016/j.jaci.2016.08.048] [PMID: 27773852]

[40] Jamnongsong S, Kueanjinda P, Buraphat P, *et al.* Comprehensive drug response profiling and pan-omic analysis identified therapeutic candidates and prognostic biomarkers for Asian cholangiocarcinoma. iScience 2022; 25(10): 105182.
[http://dx.doi.org/10.1016/j.isci.2022.105182] [PMID: 36248745]

[41] Liu Z, Zhao Y, Kong P, *et al.* Integrated multi-omics profiling yields a clinically relevant molecular classification for esophageal squamous cell carcinoma. Cancer Cell 2023; 41(1): 181-195.e9.
[http://dx.doi.org/10.1016/j.ccell.2022.12.004] [PMID: 36584672]

[42] Lin Z, Knutson KA, Pan W. Leveraging omics data to boost the power of genome-wide association studies. HGG adv 2022; 3(4): 100144.
[http://dx.doi.org/10.1016/j.xhgg.2022.100144]

[43] Chauhan MZ, Arcuri J, Park KK, *et al.* Multi-Omic analyses of growth cones at different developmental stages provides insight into pathways in adult neuroregeneration. iScience 2020; 23(2): 100836.
[http://dx.doi.org/10.1016/j.isci.2020.100836] [PMID: 32058951]

[44]   Advani D, Kumar P. Deciphering the molecular mechanism and crosstalk between Parkinson's disease and breast cancer through multi-omics and drug repurposing approach. Neuropeptides 2022; 96: 102283.
[http://dx.doi.org/10.1016/j.npep.2022.102283] [PMID: 35994781]

[45]   Razali K, Algantri K, Loh SP, Cheng SH, Mohamed W. Integrating nutriepigenomics in Parkinson's disease management: New promising strategy in the omics era. IBRO Neurosci Rep 2022; 13: 364-72.

[46]   Hu C, Ke CJ, Wu C. Identification of biomarkers for early diagnosis of Parkinson's disease by multi-omics joint analysis. Saudi J Biol Sci 2020; 27(8): 2082-8.
[http://dx.doi.org/10.1016/j.sjbs.2020.04.012] [PMID: 32714032]

[47]   Gulfidan G, Soylu M, Demirel D, *et al.* Systems biomarkers for papillary thyroid cancer prognosis and treatment through multi-omics networks. Arch Biochem Biophys 2022; 715: 109085.
[http://dx.doi.org/10.1016/j.abb.2021.109085] [PMID: 34800440]

[48]   Wang C, Lu T, Xu R, Luo S, Zhao J, Zhang L. Multi-omics analysis to identify lung squamous carcinoma lactate metabolism-related subtypes and establish related index to predict prognosis and guide immunotherapy. Comput Struct Biotechnol J 2022; 20: 4756-70.
[http://dx.doi.org/10.1016/j.csbj.2022.08.067] [PMID: 36147667]

[49]   Wu J, Vodovotz Y, Abdelhamid S, *et al.* Multi-omic analysis in injured humans: Patterns align with outcomes and treatment responses. Cell Rep Med 2021; 2(12): 100478.
[http://dx.doi.org/10.1016/j.xcrm.2021.100478] [PMID: 35028617]

[50]   Pan X, Burgman B, Wu E, Huang JH, Sahni N, Stephen Yi S. i-Modern: Integrated multi-omics network model identifies potential therapeutic targets in glioma by deep learning with interpretability. Comput Struct Biotechnol J 2022; 20: 3511-21.
[http://dx.doi.org/10.1016/j.csbj.2022.06.058] [PMID: 35860408]

[51]   Yang S, Qian L, Li Z, *et al.* Integrated multi-omics landscape of liver metastases. J Gastroenterol 2022; S0016-5085(22): 01350-6.

[52]   van der Zwet JCG, Cordo' V, Canté-Barrett K, Meijerink JPP. Multi-omic approaches to improve outcome for T-cell acute lymphoblastic leukemia patients. Adv Biol Regul 2019; 74: 100647.
[http://dx.doi.org/10.1016/j.jbior.2019.100647] [PMID: 31523030]

[53]   Passaro D, Garcia-Albornoz M, Diana G, *et al.* Integrated OMICs unveil the bone-marrow microenvironment in human leukemia. Cell Rep 2021; 35(6): 109119.
[http://dx.doi.org/10.1016/j.celrep.2021.109119] [PMID: 33979628]

[54]   Ulrich-Merzenich G, Zeitler H, Jobst D, Panek D, Vetter H, Wagner H. Application of the "-Omic-" technologies in phytomedicine. Phytomedicine 2007; 14(1): 70-82.
[http://dx.doi.org/10.1016/j.phymed.2006.11.011] [PMID: 17188482]

[55]   Lee M, Kim PJ, Joe H, Kim HG. Gene-centric multi-omics integration with convolutional encoders for cancer drug response prediction. Comput Biol Med 2022; 151(Pt A): 106192.
[http://dx.doi.org/10.1016/j.compbiomed.2022.106192]

[56]   Jiang J, Mathijs K, Timmermans L, *et al.* Omics-based identification of the combined effects of idiosyncratic drugs and inflammatory cytokines on the development of drug-induced liver injury. Toxicol Appl Pharmacol 2017; 332: 100-8.
[http://dx.doi.org/10.1016/j.taap.2017.07.014] [PMID: 28733206]

[57]   Rowida MO, Farid AB, Amal AG. An emerging flavone glycoside from *Phyllanthus emblica* L. as promiscuous enzyme inhibitor and potential therapeutic in chronic diseases. S Afr J Bot 2022; 153: 290-6.

[58]   Shi J, Qiu H, Xu Q, *et al.* Integrated multi-omics analyses reveal effects of empagliflozin on intestinal homeostasis in high-fat-diet mice. iScience 2023; 26(1): 105816.
[http://dx.doi.org/10.1016/j.isci.2022.105816] [PMID: 36636340]

[59] Zhao J, Wei W, Yan H, *et al.* Assessing capreomycin resistance on tlyA deficient and point mutation (G695A) *Mycobacterium tuberculosis* strains using multi-omics analysis. Int J Med Microbiol 2019; 309(7): 151323.
[http://dx.doi.org/10.1016/j.ijmm.2019.06.003] [PMID: 31279617]

[60] Song Y, Zhang C, Lei H, *et al.* Characterization of triclosan-induced hepatotoxicity and triclocarban-triggered enterotoxicity in mice by multiple omics screening. Sci Total Environ 2022; 838(Pt 4): 156570.
[http://dx.doi.org/10.1016/j.scitotenv.2022.156570] [PMID: 35690209]

[61] Wilmes A, Limonciel A, Aschauer L, *et al.* Application of integrated transcriptomic, proteomic and metabolomic profiling for the delineation of mechanisms of drug induced cell stress. J Proteomics 2013; 79: 180-94.
[http://dx.doi.org/10.1016/j.jprot.2012.11.022] [PMID: 23238060]

[62] Rahman F, Mahmood TB, Amin A, *et al.* A multi omics approach to reveal the key evidence of GDF10 as a novel therapeutic biomarker for breast cancer. Inform Medic Unlock 2020; 21(10): 100463.
[http://dx.doi.org/10.1016/j.imu.2020.100463]

[63] Gonçalves E, Poulos RC, Cai Z, *et al.* Pan-cancer proteomic map of 949 human cell lines. Cancer Cell 2022; 40(8): 835-849.e8.
[http://dx.doi.org/10.1016/j.ccell.2022.06.010] [PMID: 35839778]

[64] Kabir F, Apu MNH. Multi-omics analysis predicts fibronectin 1 as a prognostic biomarker in glioblastoma multiforme. Genomics 2022; 114(3): 110378.
[http://dx.doi.org/10.1016/j.ygeno.2022.110378] [PMID: 35513291]

[65] Birrell GW, Challis MP, De Paoli A, *et al.* Multi-omic characterization of the mode of action of a potent new antimalarial compound, JPC-3210, against *Plasmodium falciparum.* Mol Cell Proteomics 2020; 19(2): 308-25.
[http://dx.doi.org/10.1074/mcp.RA119.001797] [PMID: 31836637]

[66] Kappel BA, De Angelis L, Heiser M, *et al.* Cross-omics analysis revealed gut microbiome-related metabolic pathways underlying atherosclerosis development after antibiotics treatment. Mol Metab 2020; 36: 100976.
[http://dx.doi.org/10.1016/j.molmet.2020.100976] [PMID: 32251665]

[67] Mobasheri A, Kapoor M, Ali SA, Lang A, Madry H. The future of deep phenotyping in osteoarthritis: How can high throughput omics technologies advance our understanding of the cellular and molecular taxonomy of the disease? Osteoarthritis and Cartilage Open 2021; 3(4): 100144.
[http://dx.doi.org/10.1016/j.ocarto.2021.100144] [PMID: 36474763]

[68] Li R, Zhou W. Multi-omics analysis to screen potential therapeutic biomarkers for anti-cancer compounds. Heliyon 2022; 8(9): e09616.
[http://dx.doi.org/10.1016/j.heliyon.2022.e09616] [PMID: 36091949]

[69] Ye J, Huang F, Zeng H, *et al.* Multi-omics and network pharmacology study reveals the effects of Dengzhan Shengmai capsule against neuroinflammatory injury and thrombosis induced by ischemic stroke. J Ethnopharmacol 2023; 305: 116092.
[http://dx.doi.org/10.1016/j.jep.2022.116092] [PMID: 36587875]

[70] Zhou H, Lin B, Yang J, *et al.* Analysis of the mechanism of Buyang Huanwu Decoction against cerebral ischemia-reperfusion by multi-omics. J Ethnopharmacol 2023; 305: 116112.
[http://dx.doi.org/10.1016/j.jep.2022.116112] [PMID: 36581164]

[71] Wang H, Hou Y, Ma X, *et al.* Multi-omics analysis reveals the mechanisms of action and therapeutic regimens of traditional Chinese medicine, Bufei Jianpi granules: Implication for COPD drug discovery. Phytomedicine 2022; 98: 153963.
[http://dx.doi.org/10.1016/j.phymed.2022.153963] [PMID: 35121390]

[72]   Muthuramalingam P, Akassh S, Rithiga SB, *et al.* Integrated omics profiling and network pharmacology uncovers the prognostic genes and multi-targeted therapeutic bioactives to combat lung cancer. Eur J Pharmacol 2023; 940: 175479.
[http://dx.doi.org/10.1016/j.ejphar.2022.175479] [PMID: 36566006]

[73]   Xu C, Liu D, Zhang L, *et al.* AutoOmics: New multimodal approach for multi-omics research. Artif Intellig Life Sci 2021; 1: 100012.
[http://dx.doi.org/10.1016/j.ailsci.2021.100012]

[74]   Reska D, Czajkowski M, Jurczuk K, *et al.* Integration of solutions and services for multi-omics data analysis towards personalized medicine. Biocybern Biomed Eng 2021; 41(4): 1646-63.
[http://dx.doi.org/10.1016/j.bbe.2021.10.005]

[75]   Lin Z, He Y, Qiu C, *et al.* A multi-omics signature to predict the prognosis of invasive ductal carcinoma of the breast. Comput Biol Med 2022; 151(Pt A): 106291.
[http://dx.doi.org/10.1016/j.compbiomed.2022.106291]

[76]   Zhao Z, Wang S, Zucknick M, Aittokallio T. Tissue-specific identification of multi-omics features for pan-cancer drug response prediction. iScience 2022; 25(8): 104767.
[http://dx.doi.org/10.1016/j.isci.2022.104767] [PMID: 35992090]

[77]   Zhan C, Zhang Y, Liu X, *et al.* MIKB: A manually curated and comprehensive knowledge base for myocardial infarction. Comput Struct Biotechnol J 2021; 19: 6098-107.
[http://dx.doi.org/10.1016/j.csbj.2021.11.011] [PMID: 34900127]

[78]   Namba S, Konuma T, Wu KH, Zhou W, Okada Y. A practical guideline of genomics-driven drug discovery in the era of global biobank meta-analysis. Cell Genomics 2022; 2(10): 100190.
[http://dx.doi.org/10.1016/j.xgen.2022.100190] [PMID: 36778001]

[79]   Johnson BE, Creason AL, Stommel JM, *et al.* An omic and multidimensional spatial atlas from serial biopsies of an evolving metastatic breast cancer. Cell Rep Med 2022; 3(2): 100525.
[http://dx.doi.org/10.1016/j.xcrm.2022.100525] [PMID: 35243422]

[80]   Sienkiewicz K, Chen J, Chatrath A, *et al.* Detecting molecular subtypes from multi-omics datasets using SUMO. Cell Reports Methods 2022; 2(1): 100152.
[http://dx.doi.org/10.1016/j.crmeth.2021.100152] [PMID: 35211690]

[81]   Belshan M, Holbrook A, George JW, *et al.* Discovery of candidate HIV-1 latency biomarkers using an OMICs approach. Virology 2021; 558: 86-95.
[http://dx.doi.org/10.1016/j.virol.2021.03.003] [PMID: 33735754]

[82]   Li J, Lim RG, Kaye JA, *et al.* An integrated multi-omic analysis of iPSC-derived motor neurons from C9ORF72 ALS patients. iScience 2021; 24(11): 103221.
[http://dx.doi.org/10.1016/j.isci.2021.103221] [PMID: 34746695]

[83]   Zhang F, Xiao X, Hao J, Wang S, Wen Y, Guo X. CPAS: A trans-omics pathway analysis tool for jointly analyzing DNA copy number variations and mRNA expression profiles data. J Biomed Inform 2015; 53: 363-6.
[http://dx.doi.org/10.1016/j.jbi.2014.12.012] [PMID: 25546614]

[84]   Lu X, Hackman GL, Saha A, *et al.* Metabolomics-based phenotypic screens for evaluation of drug synergy *via* direct-infusion mass spectrometry. iScience 2022; 25(5): 104221.
[http://dx.doi.org/10.1016/j.isci.2022.104221] [PMID: 35494234]

[85]   Doostparast Torshizi A, Duan J, Wang K. Cell-type-specific proteogenomic signal diffusion for integrating Multi-Omics data predicts novel schizophrenia risk genes. Patterns 2020; 1(6): 100091.
[http://dx.doi.org/10.1016/j.patter.2020.100091] [PMID: 32984858]

[86]   Casey F, Negi S, Zhu J, *et al.* OmicsView: Omics data analysis through interactive visual analytics. Comput Struct Biotechnol J 2022; 20: 1277-85.
[http://dx.doi.org/10.1016/j.csbj.2022.02.022] [PMID: 35356547]

[87]   Meng C, Basunia A, Peters B, Gholami AM, Kuster B, Culhane AC. MOGSA: Integrative single sample gene-set analysis of multiple omics data. Mol Cell Proteomics 2019; 18(8) (1): S153-68.
[http://dx.doi.org/10.1074/mcp.TIR118.001251] [PMID: 31243065]

[88]   Song X, Ji J, Gleason KJ, *et al.* Insights into Impact of DNA copy number alteration and methylation on the proteogenomic landscape of human ovarian cancer *via* a multi-omics integrative analysis. Mol Cell Proteomics 2019; 18(8) (1): S52-65.
[http://dx.doi.org/10.1074/mcp.RA118.001220] [PMID: 31227599]

[89]   Zhang L, Liu M, Zhang Z, Chen D, Chen G, Liu M. Machine learning based identification of hub genes in renal clear cell carcinoma using multi-omics data. Methods 2022; 207: 110-7.
[http://dx.doi.org/10.1016/j.ymeth.2022.09.008] [PMID: 36179770]

[90]   Wu T, Hu E, Xu S, *et al.* ClusterProfiler 4.0: A universal enrichment tool for interpreting omics data. Innovation 2021; 2(3): 100141.
[http://dx.doi.org/10.1016/j.xinn.2021.100141] [PMID: 34557778]

[91]   Allendes Osorio RS, Nyström-Persson JT, Nojima Y, Kosugi Y, Mizuguchi K, Natsume-Kitatani Y. Panomicon: A web-based environment for interactive, visual analysis of multi-omics data. Heliyon 2020; 6(8): e04618.
[http://dx.doi.org/10.1016/j.heliyon.2020.e04618] [PMID: 32904262]

[92]   Hameed Y, Ejaz S. Integrative analysis of multi-omics data highlighted TP53 as a potential diagnostic and prognostic biomarker of survival in breast invasive carcinoma patients. Comput Biol Chem 2021; 92: 107457.
[http://dx.doi.org/10.1016/j.compbiolchem.2021.107457] [PMID: 33610131]

[93]   Verheijen MCT, Meier MJ, Asensio JO, *et al.* R-ODAF: Omics data analysis framework for regulatory application. Regul Toxicol Pharmacol 2022; 131: 105143.
[http://dx.doi.org/10.1016/j.yrtph.2022.105143] [PMID: 35247516]

[94]   Kel AE, Stegmaier P, Valeev T, *et al.* Multi-omics "upstream analysis" of regulatory genomic regions helps identifying targets against methotrexate resistance of colon cancer. EuPA Open Proteom 2016; 13: 1-13.
[http://dx.doi.org/10.1016/j.euprot.2016.09.002] [PMID: 29900117]

CHAPTER 2

Utilizing *in silico* Methods in New Drug Design

Olivia Pérez-Valera[1], Yuri Córdoba-Campo[2], Rafael Torres-Martínez[3], Yesica R. Cruz-Martínez[1,4] and Israel Valencia Quiroz[5,*]

[1] *Institute of Chemistry, National Autonomous University of Mexico (UNAM), Mexico City, 04510, Mexico*

[2] *Manuela Beltran University, Bucaramanga, Colombia*

[3] *Chemical Ecology and Agroecology Laboratory, Research Institute for Ecosystems and Sustainability, National Autonomous University of Mexico (UNAM), Morelia, Michoacan, Mexico*

[4] *Natural Products Bioactivity Laboratory, UBIPRO, Superior Studies Faculty (FES)-Iztacala, National Autonomous University of México (UNAM), Tlalnepantla de Baz, México State, 54090, México*

[5] *Phytochemistry Laboratory, UBIPRO, Superior Studies Faculty (FES)-Iztacala, National Autonomous University of México (UNAM), Tlalnepantla de Baz, México State, 54090, México*

Abstract: The current chapter offers a highly informative and enlightening overview of the practical implementation of molecular docking in the field of biotechnology, with a specific focus on drug discovery for a variety of ailments. Molecular docking, an incredibly powerful computational methodology, has increasingly been utilized as an essential instrument in the elucidation of drug-receptor interactions, providing invaluable insights into the process of designing drugs. This chapter delves into the fundamentals of molecular docking algorithms, offering a comprehensive understanding of their theoretical underpinnings, methodologies, and typical applications. Furthermore, this chapter elaborates on how this method is used to predict the binding affinity and orientation of potential small-molecule therapeutics to their protein targets, emphasizing the crucial role that molecular docking plays in the quest for new medications to treat various diseases. By presenting case studies across a range of diseases, this chapter effectively demonstrates the remarkable versatility of molecular docking in advancing our knowledge of disease pathogenesis and therapeutic interventions. In addition, specific diseases and their corresponding drugs are carefully examined, along with an in-depth review of molecular docking studies performed on these drugs. This detailed exploration serves as a robust foundation for researchers seeking to understand the utility of molecular docking in the development of more effective, targeted therapeutics. This chapter thus positions molecular docking as an indispensable tool in the field of biotechnology, propelling drug discovery into a new era of precision and efficiency. Overall, this chapter presents a comprehensive and inf-

---

* **Corresponding author Israel Valencia Quiroz:** Phytochemistry Laboratory, UBIPRO, Superior Studies Faculty (FES)-Iztacala, National Autonomous University of México (UNAM), Tlalnepantla de Baz, México State, 54090, México; Tel: +525556231136; E-mail: israelv@unam.mx

ormative overview of the diverse applications of molecular docking in biotechnology, providing an essential resource for researchers in the field.

Keywords: Biotechnology, Binding affinity, Drug discovery, Drug-receptor interactions, Disease pathogenesis, Molecular docking, Molecular docking algorithms, Protein targets, Precision medicine, Therapeutic interventions.

INTRODUCTION

With its broad range of applications in the areas of drug design and target molecule identification, molecular docking has become an essential tool for the creation of novel treatments for a variety of diseases, including but not limited to infectious diseases, neurodegenerative disorders, and chronic illnesses. To carefully assess the binding affinities between the possible pharmacological ligands and the matching target molecules in a three-dimensional space, this advanced methodology makes use of the power of computational algorithms. The ultimate goal is to get the two entities to fit together as well as possible, opening the door for the successful creation of novel therapeutic approaches. A potent computational method for structure-based drug design is molecular docking. It is employed to predict the interactions of tiny compounds with protein targets, such as prospective medicines. With the help of this method, the computational prediction of protein-ligand binding entails determining the binding mode first and then assessing how strong the protein-ligand interactions are. These interactions are directly correlated with ligand binding affinities, which increases the possibility of successful drug development. Numerous studies have shown that molecular docking simulations are becoming a common method for creating new medications [1].

Based on the idea of free energy, molecular docking determines the stability of a system by estimating the energy required to bind a tiny molecule (ligand) to a protein (target). It consists of two primary phases: a search process in which a number of potential ligand-target orientations are investigated to locate a stable one and a scoring process in which the orientations are ranked according to energy to determine the most likely ligand-protein configuration [2]. The utilization of state-of-the-art technology has exhibited significant triumph in hastening the revelation of pharmaceuticals for various ailments, specifically in the realm of cancer treatment. Through the combination of molecular docking and *in vitro* investigations, the procedure of identifying potential drugs to combat cancer has been notably hastened [3]. In addition to the aforementioned information, it is imperative to note that the development of new and innovative agents that combat the HIV virus has undergone substantial improvement *via* the utilization of an integrated approach. This strategy includes a wide range of methods, including molecular docking studies, virtual screening, 3D-QSAR (quantitative structure-activity relationship), and pharmacophore modeling. In a deliberate attempt to increase the effectiveness and potency of anti-HIV drugs, many techniques have been used. Recognizing the importance of these strategies in the creation of new HIV therapy drugs is crucial [4]. Similarly, molecular docking has shown efficacy in the domain of malaria research. Numerous studies have reported the identification of diverse molecules using this method, further consolidating its utility in drug discovery [5]. These findings reinforce the pivotal role molecular docking plays in biotechnology, heralding a new era of drug development for these, and potentially, numerous other diseases. Additionally, it is being used to create new medicine delivery systems, antibiotics, and vaccines. One of the crucial processes in drug creation is now the identification of therapeutic candidates against infectious diseases using a molecular docking technique, Because it is used to screen virtual libraries of drug-like molecules to identify potential leads for further drug development, the prediction of the ability of small molecules to bind to proteins has unique practical significance [6] (Fig. 1).

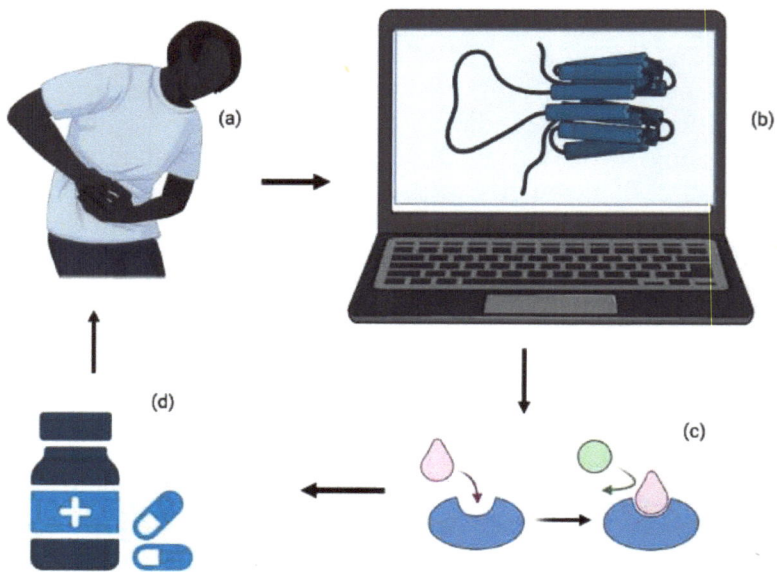

Fig. (1). Overview of the Molecular Docking-Based Drug Development Process: This flowchart illustrates a typical drug development process and highlights the critical function of molecular docking. (**a**) Determining the disease that will be targeted. (**b**) Performing molecular docking analysis to determine the binding affinity between prospective medication compounds and the identified target. (**c**) Conducting *in vitro* and *in vivo* experimental tests to confirm the outcomes of the molecular docking. (**d**) Further developing the novel medication after successful validation, and then authorizing it for therapeutic use.

A key step in the process of finding new plant-based drugs is molecular docking. To find effective novel medications, research in this area is currently concentrated on analyzing libraries of natural compounds against a particular target. Exploring natural sources for potential treatment medicines against SARS-CoV-2 is a striking example of this strategy [7]. One of the many uses for molecular docking in drug development is the virtual screening of a hundred compounds from libraries to select the one that has a certain bioactivity. Another newest application of this informatic tool is in the prediction of protein targets where a single ligand is docking with multiple receptors to find the binding to potential cavities in these targets [8].

Review of the Different Docking Algorithms, Including their Strengths and Limitations

For the development of novel medications for chronic illnesses, neurological diseases, and infectious diseases, molecular docking is a potent technique. It involves the use of computer algorithms to calculate the binding affinity of potential drug ligands and target molecules in three-dimensional space and find the best possible fit. Common algorithms used in molecular docking include Fast Fourier Transform (FFT) to calculate the correlation between the shape of the ligand and the receptor [9] and shape-based, shape-independent, energy-based, and evolutionary methods. FFT is suitable for fast and efficient searches, while shape-independent algorithms are more accurate and require a more complex scoring function. The Fast Fourier Transform (FFT) approach, which is crucially important in detecting the correlation between the shape of the ligand and the receptor, is one of the frequently used computational methodologies in molecular docking. The FFT algorithm is a key tool used in molecular docking to precisely compute the correlation between the ligand and the receptor, which is necessary to discover the best docking configuration for the specific chemical system [9].

The FFT algorithm is widely acknowledged for its remarkable swiftness and effectiveness in the execution of searches. Nevertheless, it is crucial to note that this process is largely contingent upon the configuration and form of the subject in question. In contrast to other algorithms, shape-independent algorithms are distinguished by their precise nature. These algorithms, although more intricate and requiring a more elaborate scoring function, provide an increased level of accuracy. Additional methods that are commonly utilized comprise energy-based and evolutionary approaches, each contributing their distinct strengths to the process of molecular docking.

Shape-independent algorithms are a type of computational methodology employed in molecular docking that prioritizes the energetic and chemical aspects of the interaction between two molecules over their geometric shapes. The algorithms allow for more accurate prediction of molecule binding and interactions. They consider a number of things, including solvation effects, hydrogen bonds, Van der Waals, and electrostatic forces. However, it is important to take into account the wider scope of parameters when considering these algorithms, as they typically require more complex scoring functions to effectively evaluate and rank potential docking positions. As an example, certain software applications implement a specialized methodology that thoroughly explores the complete conformational flexibility of the ligand while simultaneously maintaining a limited degree of flexibility in the protein [10]. The intricacies inherent in the aforementioned complexity of the molecular interactions lead to meticulous and all-encompassing scrutiny, thereby augmenting the strength and resilience of the docking prediction.

Shape-based algorithms are computational methodologies that focus on the geometric complementarity between two molecules, commonly a protein and a ligand. Their primary objective is to predict the binding and interactions of these molecules. These algorithms typically determine the optimal fitting pose of the ligand within the binding site of the protein based on the shapes of the respective entities. The fundamental assumption here is that a suitable geometric match indicates a favorable interaction. Compared to their shape-independent counterparts, these shape-based algorithms are usually faster and less computationally demanding [11]. However, a possible limitation of these methods is that they may not always take into account the complete range of intermolecular forces that affect binding affinity. These forces, which can be extremely important in influencing the stability of the protein-ligand complex, include solvation effects and electrostatic interactions. Shape-based algorithms utilized in molecular docking represent computational methodologies that focus on the geometric complementarity between two molecules. These two molecules are usually a protein and a ligand, and these algorithms are utilized to predict the interaction and binding between them. By examining the morphologies of the relevant entities, they may determine the best-fitting position of the ligand within the binding site of the protein. The underlying presumption of these algorithms is that a good geometric match denotes a positive interaction between the protein and the ligand.

A shape comparison program called ROCS has been used to find new scaffolds for small molecule inhibitors of a protein-protein interaction. This program evaluates the shape of a molecule, which is identified through high-throughput screening, and compares it with other molecules to identify similar shapes that may also exhibit inhibitory activity [12].

Energy-based algorithms utilized in molecular docking are a class of computational methodologies that are tailored to anticipate the binding and interactions that transpire between two molecules, typically a protein and a ligand. Their fundamental principle is to evaluate the intermolecular interaction energies that come into play during the binding process. These algorithms consider a wide variety of forces, including hydrogen bonds, Van der Waals forces, electrostatic interactions, and solvation effects. Assessing the binding affinity between the molecules is the ultimate goal. The main goal of energy-based algorithms is to achieve minimal interaction energy, which is a sign of a more advantageous binding conformation, by optimizing the position of the ligand within the binding site of the protein. This is accomplished by using Newton's equations of motion solutions, which are used to calculate the energy of a certain ligand configuration in relation to a target protein. The effectiveness of these algorithms is based on their capacity to anticipate the energy of interaction between the two molecules with a high degree of accuracy, finally resulting in a better comprehension of the binding process. The energy is then used to predict and arrange the intricate molecular structures that come from the interaction between the ligand and the protein, which are obtained through a series of complex computations [13]. In essence, it can be inferred that the configurations that have lower interaction energies are perceived as more probable and therefore hold a higher rank in the docking results.

Molecular docking is a computational technique that utilizes evolutionary algorithms such as genetic algorithms, genetic programming, or swarm intelligence to predict the interactions and binding between two molecules, typically a protein and a ligand. To find the ideal binding posture within the binding site of the protein, these algorithms repeatedly generate, mutate, and recombine possible ligand conformations. This mimics the natural selection process. The use of a genetic algorithm in software to explore the full conformational flexibility of the ligand while preserving only partial flexibility of the protein is referred to as the "evolutionary technique". The genetic algorithm, an optimization technique inspired by the natural selection process, involves a population of candidate ligand conformations that could potentially bind to the protein, selecting the best conformations based on their fit to the binding site [10].

Unquestionably, the study and interpretation of many therapeutic medicines for a variety of diseases have been greatly aided by molecular docking, a computer technique used to anticipate the binding mechanism of ligands to their particular protein targets. In the discussion that follows, we will try to elaborate fully and thoroughly on a few key examples to highlight the numerous and varied ways in which this ground-breaking methodology has been used in the field of drug discovery and development.

INFECTIOUS DISEASES (TUBERCULOSIS)

Tuberculosis, a highly contagious respiratory bacterial infection caused by *Mycobacterium tuberculosis*, has emerged as an utmost concern to global health. With an alarming rate of nearly nine million newly reported cases each year, it has become a major health crisis worldwide [14]. Notwithstanding the fact that there are pharmaceutical agents accessible to manage this infection, tuberculosis, still continues to be a pernicious malady, inflicting the demise of around two million individuals every year. The microorganism predominantly infiltrates the human body through the respiratory system, resulting in malady in merely a proportion of those who have been exposed to it [15]. Significant strides have been made in the battle against tuberculosis, with the advent of innovative treatments. Notably, the American Food and Drug Administration (FDA) approved bedaquiline fumarate, an oral diarylquinoline drug, in December 2012. This pharmaceutical agent boasts a revolutionary mode of action against the pathogen *Mycobacterium tuberculosis*, as it disrupts ATP synthetase. It is of utmost importance to recognize that this landmark drug has heralded the emergence of a new class of anti-TB medications, the first to be introduced in a span of more than four decades [16]. Notwithstanding the tremendous strides that have been made thus far in the realm of tuberculosis, it is imperative that an unceasing effort to conduct extensive research and foster the creation of avant-garde treatment modalities be pursued in view of the intricacy of this disease. This particular approach has proven to be exceedingly efficacious in regard to managing voluminous datasets and maneuvering intricate search domains, wherein conventional optimization techniques may encounter difficulties. By emulating the fundamental tenets of evolution, these algorithms can gradually enhance potential solutions across numerous generations to ascertain an optimal or nearly optimal resolution.

Tuberculosis, has been a subject of considerable concern to the global health community. The alarming frequency of new cases of this disease reported every year, which hovers around the staggering figure of approximately nine million across diverse regions and geographies of the world, has made it an issue of paramount significance and urgency [14]. Despite the fact that there are existing pharmaceutical agents that are capable of treating tuberculosis, it still persists as a fatal disease and accounts for the loss of approximately two million lives on a yearly basis. The bacillus, which is the causative agent of this malady, primarily invades the human body through the respiratory tract, thereby inducing pathological manifestations in only a fraction of the total number of individuals who have been exposed to it [15].

Despite the enormous advancements in tuberculosis diagnosis and treatment, the complex nature of this illness necessitates a constant search for new and cutting-edge therapeutic approaches. In light of the persistent challenges posed by this debilitating disease, it is imperative that the scientific community continue to engage in extensive research and development aimed at unraveling the intricate puzzle of tuberculosis. In the current work, a molecular docking strategy was used. The c-subunit of the three-dimensional (3D) structure of ATPase from *Mycobacterium tuberculosis* was initially carefully modeled and then improved. This was achieved through the proficient utilization of the Protein Preparation Wizard, which is an innovative software tool developed by Schrodinger. It is noteworthy that the binding site of the aforementioned model was enclosed by a 3D grid, which was meticulously constructed. Then, utilizing Glide SP docking software, the indicated inhibitors, including bedaquiline, were docked [17].

Case Study: Bedaquiline

The screening of many compounds is a necessary step in the difficult and convoluted process of drug discovery using molecular docking techniques. In this instance, bedaquiline was found using molecular docking screening of a library of diarylquinoline molecules. Using this method led to the discovery of a particular diarylquinoline molecule, R207910, which showed significant *in vitro* inhibition of both drug-sensitive and drug-resistant *Mycobacterium tuberculosis*. It is noteworthy that this particular chemical significantly inhibited the ATP synthase of *Mycobacterium tuberculosis*, suggesting that it may be a good candidate for the treatment of tuberculosis [18].

Clinical Trials and Approval

Bedaquiline underwent clinical studies after it was discovered, and these studies showed that it was successful in treating drug-resistant TB. As a result, bedaquiline quickly received approval from the US Food and Drug Administration (FDA) in 2012 to treat multidrug-resistant tuberculosis [16].

Real-World Impact on Disease Treatment

The inclusion of bedaquiline, a diarylquinoline antibiotic, in a comprehensive therapeutic regimen for patients afflicted with multidrug-resistant tuberculosis (MDR-TB) was well-tolerated, with minimal adverse effects, and furthermore, resulted in favorable clinical outcomes among a particularly relevant population of patients with MDR-TB [16]. This illustrates the significant real-world impact of this groundbreaking drug on TB treatment.

CANCER (RENAL CELL CARCINOMA AND HEPATOCELLULAR CARCINOMA)

Disease Characteristics

Renal cell carcinoma, a pathological condition that is mainly characterized by the malignant growth of epithelial cells in the kidneys, is the most prevalent histological subtype of kidney cancer. In fact, this type of cancer accounts for approximately 90% of all clinically determined instances of kidney cancer globally [19]. The most common form of primary liver cancer and the fourth leading cause of cancer-related deaths worldwide is hepatocellular carcinoma (HCC), a malignancy that develops from hepatocytes, the basic functioning cells of the liver [20]. Renal cell carcinoma (RCC) and hepatocellular carcinoma (HCC) are two cancers that have the potential to develop without any symptoms in the early stages. However, as the course of either of these pathological conditions progresses, a plethora of vexatious symptoms can arise, encompassing but not confined to anguish, reduction in body mass, and other diverse corporeal signs of the ailment. It should be noted that notable risk factors associated with RCC include smoking, corpulence, and hypertension, whereas HCC is frequently associated with chronic liver diseases such as hepatitis B and C and cirrhosis.

Molecular Targets and their Role in Disease

One molecular target that both renal cell carcinoma (RCC) and hepatocellular carcinoma (HCC) share is none other than the vascular endothelial growth factor receptor (VEGFR), according to a thorough examination of the molecular targets and their relative roles in disease. This receptor, which is widely expressed in both RCC and HCC, regulates angiogenesis, a vital biological process that involves the growth of new blood vessels that are critical for supplying the growing tumor with nutrients [21]. Inhibition of the vascular endothelial growth factor receptor (VEGFR) presents a highly attractive and viable therapeutic target for efficient cancer treatment since it has the ability to greatly slow tumor development and metastasis.

A method of molecular docking was used to find compounds that could prevent the vascular endothelial growth factor receptor (VEGFR) from performing its function. This technique involved the scrutiny of interactions that occurred between the protein and ligand, wherein a scoring function was utilized to grade the potential inhibitory elements. The compounds that garnered the highest scores were subsequently subjected to a series of structural alterations, the objective of which was to optimize their capability to impede VEGFR [22]. This particular methodology combines the fields of computational biology and chemical synthesis to create precise and targeted treatments for various types of cancers.

Utilizing molecular docking is a crucial stage in the drug development process since it allows sorafenib to be recognized as a very effective VEGFR inhibitor. Through extensive preclinical studies, this particular compound displayed exceptional antiangiogenic and antitumor properties, solidifying its potential as a *via*ble treatment option for various types of cancers [22]. Following its development, sorafenib was put through a variety of rigorous clinical trials, which ultimately led to its approval for the effective care of unresectable hepatocellular carcinoma (HCC) in 2007 and advanced renal cell carcinoma (RCC) in 2005 [21]. The approval of the orally administered multikinase inhibitor for these particular indications signified a groundbreaking development in targeted cancer therapy, representing a considerable step forward in the field. This novel advancement, which was the first of its kind, marked a notable milestone in the ongoing quest to combat cancer through the use of highly tailored, precision medicine approaches. The approval of this inhibitor represents a major triumph for the field of oncology, as it holds enormous promise for improving patient outcomes and reducing the burden of this devastating disease. Overall, this milestone achievement is a major milestone in the ongoing fight against cancer and underscores the importance of continued research and development in the field of targeted therapy.

Real-World Impact on Disease Treatment

The use of Sorafenib has significantly improved progression-free survival and overall survival in these patient populations, which has had a profound effect on the therapy of advanced renal cell carcinoma (RCC) and unresectable hepatocellular carcinoma (HCC) [21] [23]. Its approval has spurred the development of other VEGFR inhibitors and has broadened the application of molecular docking in the discovery of targeted cancer therapies. Inhibiting the PDGF receptor beta and VEGFR2 pathways has been observed to impede angiogenesis, hamper vascular maturation, and suppress cell proliferation, ultimately culminating in tumor regression. It is noteworthy that the FDA has granted approval to various inhibitors of VEGFR2 and PDGFRβ, including bevacizumab (Avastin), sunitinib malate, and sorafenib [24].

Molecular docking and molecular dynamics techniques were used to provide important knowledge about the interaction between sorafenib and VEGFR2. According to the results of the docking analysis, sorafenib forms important hydrogen bonds with the Asp1046, Cys919, and Glu885 residues of the VEGFR2 receptor. The findings of the molecular dynamics simulation showed that the Asp1046 hydrogen bond is the most stable bond, and it maintains its stability for the entire course of the MD simulation [25]. This particular piece of insight, which has been painstakingly determined through rigorous scientific investigation, pertains to the manner in which sorafenib binds to its target. It is our ardent belief that this information can serve as a valuable guidepost, illuminating the path toward the development of newer, more efficacious and selective inhibitors of VEGFR2.

CARDIOVASCULAR DISEASES (VENOUS THROMBOEMBOLISM AND STROKE)

Disease Characteristics

Cardiovascular illnesses have a wide range and complexity of symptoms. These conditions have numerous negative effects on the heart and blood arteries, including the occurrence of venous thromboembolism (VTE) and stroke. Deep vein thrombosis, or VTE, is a disorder where blood clots can form and travel to the lungs, which causes a potentially fatal pulmonary embolism. A stroke is an event that occurs when blood flow to the brain is interrupted, either *via* a blood clot causing an ischemic stroke or a ruptured blood vessel causing a hemorrhagic stroke. Both VTE and stroke can result in severe disability or even death. In fact, risk factors that are associated with these conditions include obesity, smoking, a sedentary lifestyle, and genetic predisposition. Therefore, it is important to recognize and address these risk factors to prevent these diseases and their devastating consequences [26].

Molecular Target and their Role in the Disease

The highly efficient, aggressive, and completely reversible direct inhibitor melagatran may firmly bind to the active site of both soluble and clot-bound α-thrombin. It is worth mentioning that melagatran shows a wide range of antithrombotic properties, including the inhibition of thrombin and platelet aggregation, as well as other marked effects, such as enhanced fibrinolysis. It is important to note that ximelagatran, which is a derivative of melagatran, does not present any pharmacodynamic effect whatsoever [27].

Even at very high dosages, the fast and competitive thrombin inhibitor melagatran, with a molecular mass of 429 Da, has been shown to be well tolerated. It also has relatively high oral bioavailability in dogs. The level of selectivity against the fibrinolytic system required to enter the clinical development phase was investigated in preclinical research. A low inhibition constant (K_i) for thrombin and an extended clotting time in coagulation assays at low concentrations demonstrated the potency of melagatran in inhibiting thrombin [28]. Empirical research has been performed to examine the binding interactions of melagatran with thrombin. These investigations have demonstrated that melagatran securely binds to the thrombin active site, generating a variety of hydrogen bonds and hydrophobic interactions with protein residues [29]. Gaining a more thorough understanding of the interaction between melagatran and thrombin may aid in the ongoing effort to create and improve thrombin inhibitors with the eventual goal of using those inhibitors to treat cardiovascular diseases.

Ximelagatran, a prodrug of melagatran, has been developed to augment its pharmacokinetic properties with dissimilarities from melagatran. It can be administered orally and is rapidly absorbed through the gastrointestinal tract. After administration, ximelagatran is converted into melagatran by the liver. In the clinical setting, ximelagatran has been used for the prevention and treatment of thromboembolic conditions, such as deep vein thrombosis and pulmonary embolism. However, because its use has been linked to liver toxicity, ximelagatran has been taken off the market in various nations. Initially, ximelagatran, an agent that directly inhibits thrombin, demonstrated comparable effectiveness and safety to the well-known anticoagulant warfarin. Nevertheless, subsequent research revealed that ximelagatran exhibited hepatotoxicity, thereby impeding its continued application [30]. In contrast to ximelagatran, dabigatran, which is also a direct thrombin inhibitor, did not exhibit discernible manifestations of hepatotoxicity when examined *via* serial liver function tests; therefore, it has potential as a comparatively safer option [30].

AUTOIMMUNE DISEASES (RHEUMATOID ARTHRITIS)

Rheumatoid arthritis (RA) is a chronic autoimmune illness that typically progresses and results in a variety of incapacitating symptoms, including excruciating pain, joint damage, and functional incapacity [31]. RA has been estimated to have affected 460 out of every 100,000 people worldwide from 1980 to 2019 [32]. Notably, smoking is a recognized environmental risk factor for rheumatoid arthritis development. In addition to cigarette smoking, there are other potential factors that could contribute to the emergence of this condition, including exposure to silica dust, mineral oils, and other airway irritants. It is particularly intriguing to note that charcoal workers have historically been diagnosed with a severe form of RA, commonly known as Caplan's syndrome [33].

Inhibitors of the JAK family, which include nonreceptor tyrosine kinases, have been demonstrated to have significant therapeutic efficacy against rheumatoid arthritis as well as other inflammatory diseases related to the molecular targets that have been identified thus far. However, the specific set of mechanisms underlying the increase in inflammatory immune responses *via* JAK inhibition still needs more development and investigation, despite the promising findings that have been produced through such interventions [34]. The current study focuses on tofacitinib, a novel oral Janus kinase inhibitor that is now being evaluated for its effectiveness in treating RA. It is also known by the trade name CP-690,550 [35]. Tofacitinib selectively inhibited JAK-1, JAK-2, and JAK-3 *in vitro* tests and revealed the functional cellular selectivity of JAK-1 and JAK-3 over JAK-2 [27]. This inhibition is thought to impede signaling through cytokines with common chains, including interleukin-2 (IL-2), IL-4, IL-7, IL-9, IL-15, and IL-21. However, as JAK2 mediates signaling *via* a variety of hematopoietic cytokines, including erythropoietin, thrombopoietin, and colony-stimulating factor receptors, it is important to note that effective JAK2 inhibition could potentially result in anemia, thrombocytopenia, and leukopenia *in vivo* [36].

A molecular docking method was used to identify tofacitinib, a potent inhibitor of Janus kinase (JAK). This method showed that tofacitinib can effectively interact with all JAKs at the ATP-binding site through electrostatic attraction, hydrogen bond formation, and most notably, Van der Waals interactions [37]. Tofacitinib was introduced in clinical trials after this finding, where its efficacy was convincingly demonstrated. The percentage of patients who met the ACR 20 response criteria after three months was considerably higher in the tofacitinib group than in the placebo group (59.8% for the 5-mg dose and 65.7% for the 10-mg dose) (26.7%). The HAQ-DI score reductions from baseline were also more pronounced in the tofacitinib groups. Despite some negative side effects being reported, including severe infections, headaches, and upper respiratory tract infections, patients reported significantly less fatigue and pain [38].

In the real world, tofacitinib has considerably affected the way diseases are treated. As the first Janus kinase inhibitor approved by the FDA, it has provided rheumatoid arthritis sufferers with a distinctive therapeutic option for inflammation. The medication, sold under the brand name Xeljanz and made by Pfizer, can be used either alone or in conjunction with methotrexate or other nonbiologic disease-modifying antirheumatic medications (DMARDs). It should not, however, be taken in conjunction with several other therapies. Upper respiratory tract infections, headaches, diarrhea, and nasopharyngeal irritation were the most frequently reported side effects during clinical studies [39]. Tofacitinib for rheumatoid arthritis was recommended for approval by the European Medicines Agency (EMA) in 2017. Additionally, the FDA authorized the extended-release formulation of tofacitinib in 2016; this formulation uses osmotic administration for once-daily treatment [40]. The FDA-approved labeling of tofacitinib, however, includes a boxed warning about the possibility of life-threatening infections, lymphomas, and other malignancies in individuals using the medication [39]. The successful development of tofacitinib has illustrated the promise of molecular docking in the investigation and creation of novel therapeutic medicines for autoimmune disorders. As our understanding of the molecular pathways underlying these diseases increases, molecular docking may

be used more frequently to identify new targets and create novel therapies that address the unmet needs of individuals suffering from these diseases.

METABOLIC DISORDERS (TYPE 2 DIABETES)

In the realm of healthy individuals, it has been empirically observed that the process of oral glucose intake activates a notably stronger insulin secretory response when juxtaposed against the intravenous glucose process, even when the resultant glycemic increments are similar. This notable phenomenon, universally recognized as the incretin effect, is markedly reduced or, in certain cases, even completely absent in individuals who suffer from type 2 diabetes [41]. Diabetes mellitus, an endocrine disorder, is a multifaceted group of metabolic conditions that are typified by the chronic elevation of blood glucose levels over a prolonged period. This condition is primarily defined by the inability of the body to produce or utilize insulin, a hormone that regulates glucose metabolism. The clinical manifestations of this condition may include polyuria, polydipsia, and polyphagia, which are characterized by frequent urination, increased thirst, and increased hunger, respectively. These symptoms are frequently accompanied by a variety of other issues that can significantly lower an individual's quality of life [42].

The incretin system, which is crucial in controlling glucose levels, is predominantly linked to the molecular targets implicated in type 2 diabetes. The postprandial spike in blood sugar is well controlled by incretins, such as glucagon-like peptide-1 (GLP-1) and glucose-dependent insulinotropic polypeptide (GIP), and thus, fasting glucose levels are reduced. The enzyme dipeptidyl peptidase-4 (DPP-4) is capable of rapidly inactivating both GLP-1 and GIP, which can considerably reduce their therapeutic efficacy in the management of type 2 diabetes. Determining the complicated interactions between the incretin system and DPP-4 is therefore critically needed to create more potent therapies for this complex metabolic condition [43].

Sitagliptin, also known as MK-0431, is an oral dipeptidyl peptidase-4 (DPP-4) inhibitor that has been specifically created and developed to treat type 2 diabetes. It demonstrates high efficacy and selectivity. The active incretin hormone concentrations of the body are increased as part of the sitagliptin mechanism of action. Sitagliptin has been demonstrated to efficiently sustain a 24-hour suppression of DPP-4 activity in individuals with type 2 diabetes. This results in a marked increase in active GLP-1 (glucagon-like peptide-1) and glucose-dependent insulinotropic polypeptide (GIP) concentrations. As a result of this increase in active hormone levels, insulin and C-peptide levels are increased, glucagon levels are decreased, and oral glucose tolerance is eventually improved [43].

An essential role in the metabolism of incretin hormones, including glucagon-like peptide-1 (GLP-1) and glucose-dependent insulinotropic polypeptide, is played by serine protease dipeptidyl peptidase-4 (DPP-4) (GIP). DPP-4 inhibitors work by delaying the activation of these incretin hormones, which in turn increases insulin secretion, decreases glucagon release, and subsequently improves glycemic control [41].

Three residues in the active site—Try226, Glu205, and Glu206—play a crucial role in the creation of three hydrogen bonds, according to molecular docking studies of the interactions between dipeptidyl peptidase-4 (DPP-4) and sitagliptin. Notably, hydrophobic interactions involve eleven more amino acids, Try547, Try667, Asn710, Val711, His740, Ser630, Ser209, Arg358, Phe357, and Val207 [44]. Sitagliptin was discovered to be a powerful and highly selective DPP-4 inhibitor by using molecular docking techniques [44]. When compared to a placebo, the substance significantly decreased fasting plasma glucose levels during the preclinical phase [43].

Due to these promising findings, clinical trials for this medicine were performed. The phase III study findings, which were presented at the American Diabetes Association meeting in June 2006, revealed that sitagliptin was well tolerated at doses of 100 mg once daily. There were no notable reports of hypoglycemia or weight gain when the medication was used alone or in combination with metformin or pioglitazone [41]. By successfully stopping the enzymatic degradation of two important incretins, glucose-dependent insulinotropic peptide (GIP) and glucagon-like peptide-1 (GLP-1), involved in glucose homeostasis, the oral and highly selective DPP-4 inhibitor sitagliptin presented a novel approach to treat type 2 diabetes. This medication effectively lowered the postprandial glucose concentration in type 2 diabetes patients by doubling active GLP-1 and GIP concentrations, enhancing insulin release, and decreasing glucagon production in a glucose-dependent manner [45]. These results were crucial in getting the medication approved in the USA for the treatment of type 2 diabetes in October 2006 [41].

Sitagliptin has been found to be very effective in the control of type 2 diabetes when used in practice. The findings from large-scale clinical trials show that this medication lowers A1C, fasting, and postprandial glucose concentrations by amounts that are clinically relevant. Additionally, it has been found to be well tolerated when used alone or in combination with metformin or pioglitazone [45]. It is important to note that despite the significant advancements in glycemic control, hypoglycemia is uncommon in all therapy groups. This result is consistent with the impact of incretins on glucose levels [45]. The successful invention and use of sitagliptin is evidence of the enormous potential of molecular docking in the search for and development of novel therapeutic medicines for metabolic diseases such as type 2 diabetes. Molecular docking is projected to play an increasingly important role in the identification of novel drug targets and the development of individualized treatment approaches as scientific knowledge and technological capabilities continue to advance. Molecular docking can surely significantly contribute to the improvement of patient outcomes and the reduction of the worldwide burden of metabolic illnesses through the coordinated efforts of researchers from several disciplines.

PSYCHIATRIC DISORDERS (DEPRESSION)

Disease Characteristics

Depression, a psychiatric disorder that is highly prevalent in society, is most notably distinguished by a significant and profound sense of isolation experienced by individuals in their adolescent years. This sense of isolation is intricately linked to five distinct clusters of symptoms: a persistent low mood and the inability to experience pleasure, otherwise known as anhedonia; sleep and appetite disturbances often accompanied by fatigue; an amplified sense of irritability and anger; a negative self-perception, which ultimately leads to feelings of hopelessness and self-doubt; and a propensity toward suicidal thoughts or inclinations [46].

Molecular Targets and their Role in the Disease

The pivotal roles played by monoamine neurotransmitters, which include serotonin, norepinephrine, and dopamine, in the pathophysiology of depression have been well established. It has been found that disruption in the monoaminergic systems within the brain is closely linked to the emergence of depressive symptoms. As a result of their critical role in the control of these neurotransmitters, the serotonin transporter (SERT), norepinephrine transporter (NET), and dopamine transporter (DAT) have been identified as major targets in the development of antidepressant medicines. These neurotransmitter levels are frequently noticeably decreased in people who are depressed. Antidepressants work primarily by inhibiting the reuptake of these neurotransmitters by their respective transporters, which increases the levels of those neurotransmitters in the brain. On this basis, the effectiveness of these drugs in alleviating the symptoms of depression has been established [47].

Molecular Docking Approach

Inhibitors of neurotransmitter transporters, which are essential for proper neural signaling, generally possess a bicyclic structure that comprises a pyridine ring and a piperidine ring. These synthetic compounds have undergone rigorous testing to evaluate their inhibitory activity against three key transporters, namely, DAT, SERT, and NET, and their binding affinity was predicted *via* molecular docking studies. It is worth noting that atypical inhibitors have displayed promising outcomes by effectively preventing the reuptake of neurotransmitters, thereby positioning them as compelling therapeutic candidates for a diverse array of neurological disorders [48].

Case Study: Vortioxetine

Vortioxetine, a kind of multimodal antidepressant, has been shown to be effective in lowering somatic symptoms that are prevalent in individuals with severe depressive disorder (MDD) [49]. Additionally, it has been shown that the safety and acceptability of vortioxetine are equivalent to those of selective serotonin reuptake inhibitor therapy [50]. However, certain patients may benefit from the particular tolerability profile of vortioxetine in terms of weight gain and sexual dysfunction [49]. Vortioxetine specifically interacts with SERT and the 5-HT1A, 5-HT1B, 5-HT3, and 5-HT7 receptors at therapeutic levels and is sold under the brand name Trintellix [51].

Drug Development Process using Molecular Docking

To explore the binding location of the chemical vortioxetine in the serotonin transporter, molecular modeling approaches were used in the current work (SERT). To promote more thorough knowledge of the binding mode by induced fit docking, homology models of human SERT (hSERT) were specifically used in various conformational states (IFDs). IFD has been frequently utilized as a computational method to forecast how a ligand will bind to a protein. IFD was used in this work to predict the binding mode of vortioxetine to hSERT. By using X-ray crystal structures of dDAT and LeuT to generate homology models of hSERT in two different conformational states, the binding mode could be predicted more accurately [52].

Clinical Trials and Approval

Vortioxetine, a medication used to treat major depressive disorder (MDD) patients, has been found to possess significant clinical efficacy in addressing depressive symptoms and cognitive deficits in both short-term and long-term clinical trials. Furthermore, it has been observed to exhibit a generally benign safety and tolerability profile, which is a crucial aspect of any medication. The unique psychopharmacological properties of vortioxetine could potentially improve clinical outcomes in MDD patient populations, as evidenced by a study [53]. The United States Food and Drug Administration (FDA) approved vortioxetine for the treatment of major depressive disorder in adults on September 30, 2013 [54].

Real-World Impact on Disease Treatment

Vortioxetine, a pharmaceutical substance, has garnered considerable attention in the medical community due to its efficacy in mitigating symptoms of depression and augmenting the overall quality of life for patients. This novel medication has a profound impact on a global scale, serving as a viable alternative treatment option for individuals afflicted with depression who have not experienced an optimal response to other conventional drugs. It is noteworthy to mention that vortioxetine has obtained regulatory approval for use in diverse countries worldwide, including the United States, Canada, European countries, and Japan [54].

The utilization of molecular docking methodologies in the exploration and advancement of vortioxetine has resulted in a crucial and noteworthy addition to the available treatment options for depression. Its unique and distinct multimodal mechanism of action confers certain benefits in comparison to traditional antidepressants, particularly with respect to the augmentation of cognitive function. In the context of ongoing research into the neurobiology of depression and other psychiatric disorders, it is anticipated that molecular docking techniques will aid in the identification of novel drug targets and the formulation of more efficacious and individualized treatment options. These scientific advances have the potential to significantly impact the well-being of millions affected by depression and other psychiatric illnesses, thereby elevating the quality of care and catalyzing the development of innovative therapeutic strategies.

GENETIC DISORDERS (CYSTIC FIBROSIS)

Mutations in the cystic fibrosis transmembrane conductance regulator (CFTR) gene, which affects the synthesis and function of the CFTR protein, cause cystic fibrosis (CF), a hereditary condition that causes slow lung deterioration and premature death. When the effectiveness of this protein is impaired, it leads to an accumulation of sticky, thick mucus in the pancreas, lungs, and other internal organs. This protein controls the passage of salt and water through cells. The said mucus clogs airways, impairing breathing and perhaps causing infections and other issues [55].

Disease Characteristics

Although many physiological organs are negatively influenced by the inherited disease cystic fibrosis (CF), the respiratory system is most typically affected. Clinical manifestations of the disorder include persistent coughing, wheezing, dyspnea, and recurrent lung infections such as pneumonia or bronchitis. Additionally, alimentary issues, including insufficient growth, malnourishment, and recurring voluminous, oily feces, are widespread. In addition to these symptoms, individuals with CF may also experience skin that tastes salty, infertility in males, and sinusitis [55].

Molecular Targets and their Role in the Disease

The transmembrane conductance regulator (CFTR) protein is a key player in controlling the flow of salt and water through cells. Mutations in the CFTR gene affect the generation and operation of the protein, disrupting this crucial function. As a result, the lungs, pancreas, and other organs develop a buildup of thick, sticky mucus. Numerous treatment targets that try to improve or compensate for the reduced function of the CFTR protein and target the mechanisms behind progressive lung disease have been identified to address this issue. For cystic fibrosis, in particular, various possible treatment targets are being investigated. These include pharmacological medicines that promote Cl- secretion through alternate Cl- channels or inhibit ENaC, chemicals that improve CFTR expression and function, and agents that target the mechanisms behind progressive lung disease. It is important to note that some of these substances, such as Moli190, the purinergic (P2Y2) agonist INS3721, and amiloride, are now being investigated in clinical trials [55]. Through these methods, it may be possible to mitigate the deleterious effects of CFTR mutations and improve the quality of life for those afflicted with this condition.

Molecular Docking Approach

In the current study, molecular docking studies were used to thoroughly investigate the docking mechanism of the prototype VX-809 and its correctors within the model F508del-CFTR (cystic fibrosis transmembrane regulator protein). It was found that VX-809 displayed an extra hydrogen bond (H-bond) between the carboxamide nitrogen atom and K1060 as a result of the folded shape of the corrector. Additionally, it was discovered that the pyridine ring made contact with W496 by interactions, whereas the terminal carboxy-substituted phenyl ring made contact with F494 and K1060 through cation contacts and stacking. The cyclopropyl component additionally displayed van der Waals interactions with W1063 and C1344 [56].

Class I corrector ligands, such as VX-809, VX-661, and C18, were used in molecular docking. A multidomain pocket near residues F374–L375 is thought to be the likely corrector binding site according to in silico molecular docking results for the complex of corrector molecules and the cryo-electron microscopy structure. Small compounds known as class I correctors, such as VX-809, VX-661, and C18, can improve the folding, stability, and trafficking of mutant CFTR proteins to the cell surface. In the context of this study, it is anticipated that these correctors bind to the CFTR protein in its closed-channel conformation. The mechanism of action of these corrector molecules can be better understood by determining their binding location, which also makes it possible to stratify patients according to their mutation-sensitive "theratype" [57].

Case Study: Lumacaftor

Lumacaftor is a pharmaceutical agent that has been employed as a remedial measure in the management of cystic fibrosis. This minuscule chemical entity has been shown to augment the expression of F508del CFTR, a commonly occurring folding mutant that is predisposed to degradation rather than being localized on the cell surface. The administration of lumacaftor is often accompanied by the use of another corrective drug, ivacaftor, to achieve an optimal therapeutic outcome. Furthermore, this combination therapy may also be of great benefit to patients who are currently undergoing potentiator therapy. Ivacaftor functions by increasing the open probability of CFTR channels, thereby enhancing protein function on the cell surface. The beneficial outcomes of the synergistic effects of combining correctors and potentiators may be especially advantageous for patients with functional CFTR mutants [58].

Drug Development Process using Molecular Docking

Lumacaftor was developed as a consequence of the application of molecular docking techniques. These techniques were used to successfully identify compounds with the ability to correct the folding and trafficking of F508del, which happens to be the most common CFTR mutation that causes cystic fibrosis [59].

Clinical Trials and Approval

The percentage predicted forced expiratory volume in 1 second (ppFEV1) compared to the baseline and a substantial 35 percent drop in pulmonary exacerbations have both been found as signs of the treatment efficacy of lumacaftor/ivacaftor (LUM/IVA). LUM/IVA therapy has a well-established track record of both short- and long-term safety and tolerability. LUM/IVA continues to be the sole oral medication in its class at this time and represents a significant advancement in the development of cystic fibrosis (CF) treatments.

Hepatic function monitoring is necessary before starting LUM/IVA medication. This should be followed by quarterly tests throughout the first year of treatment and then an annual evaluation. Common side effects that patients may suffer include upper respiratory infections, gastrointestinal problems, and increased liver enzyme levels. However, it is possible for uncommon events to trigger serious liver problems. Patients are urged to call their doctor right away if they experience any liver-related symptoms, such as skin or eye yellowing, black urine, or pain in the upper right abdomen region.

UM/IVA (Orkambi), a CFTR modulator made up of a corrector and potentiator, was approved by the FDA and the European Commission in 2016 for individuals harboring the F508del CFTR mutation. This medication is intended for those with CF who harbor the homozygous F508del CFTR mutation [60].

The investigational medicine in question was administered to 1,018 human volunteers in total after they underwent randomization procedures. According to calculations, the average baseline forced expiratory volume in one second (FEV1) was 61% of the expected value. The primary endpoint in both lumacaftor-ivacaftor dosages showed considerable improvements in the findings from both studies that were performed. Comparing the lumacaftor-ivacaftor groups to the placebo group, analyses of the pooled data showed a thirty to thirty-nine percent reduction in the frequency of pulmonary exacerbations [61].

Real-World Impact on Disease Treatment

The drug combination lumacaftor/ivacaftor (LUM/IVA) marks a significant advancement in the care of cystic fibrosis (CF) patients. It has shown excellent results in improving lung function and lowering pulmonary exacerbations in CF patients homozygous for the CFTR (cystic fibrosis transmembrane conductance regulator) mutation F508del. It is the only oral drug in its class. However, before it can become the standard of care, pharmacoeconomic statistics are needed to support its high cost. Despite this, LUM/IVA has helped CF patients harboring the F508del CFTR mutation live happier and healthier lives by providing a reliable therapy choice [60].

Molecular docking techniques were used in the generation of lumacaftor, demonstrating their potential to advance drug discovery and development and, eventually, produce more potent treatments. There are millions of individuals with cystic fibrosis and other genetic abnormalities in the world today, and ongoing research in this field may have a substantial impact on their quality of life. These developments signify an ongoing improvement in the standard of treatment, creating the groundwork for novel therapeutic approaches to be developed.

BONE-RELATED DISEASES (OSTEOPOROSIS)

A common and crippling condition called osteoporosis is a skeletal disorder characterized by a decline in bone density, which increases the risk of fractures [62]. This condition is systemic in nature, and it results from an imbalance in bone remodeling, which is an ongoing process of bone creation and resorption [63]. Across various ethnic groups, osteoporosis affects approximately 200 million people on a global scale [64]. The molecular target in the pathophysiology of osteoporosis is the lysosomal cysteine protease cathepsin K. By maintaining the equilibrium between bone production and bone resorption, this enzyme is essential for bone resorption. Inhibiting cathepsin K has therefore become a promising therapeutic approach for the management of osteoporosis [65]. Small compounds that interact with cathepsin K can be found and improved using the molecular docking technique. This technique allows for the analysis of protein-ligand interactions, ranks possible ligands according to how well they bind to the protein, and tweaks the structure of the ligands for increased potency and selectivity [66].

Case Study: Odanacatib

The utilization of molecular docking strategies that were created especially to find powerful and selective inhibitors of cathepsin K helped lead to the discovery of odanacatib. To create a highly efficient inhibitor, it was essential to optimize the interaction between odanacatib and cathepsin K during its development [66]. After odanacatib was discovered, the clinical trials for odanacatib showed that it could successfully cut bone resorption, improve bone mineral density, and lessen the risk of fractures in postmenopausal women with osteoporosis. The development of odanacatib was halted in 2016 despite these encouraging outcomes because of safety issues that surfaced during a phase III clinical trial [67]. Despite the fact that odanacatib was not given final regulatory approval, the molecular docking approach used in its development has greatly advanced our knowledge of cathepsin K as a potential therapeutic target. Future therapies for disorders connected to bones, such as osteoporosis, could be developed using this knowledge.

The creation of odanacatib demonstrates the potential advantages of using molecular docking techniques in the identification and improvement of small compounds that target cathepsin K and other enzymes. Despite the discontinuation of odanacatib due to safety concerns, the insights gained from its development may facilitate the creation of future drugs with improved safety profiles for treating osteoporosis and other bone-related diseases. The molecular docking techniques that were utilized in the development of odanacatib provide a valuable model for demonstrating how this approach can contribute to the discovery of innovative therapeutics for a wide range of health conditions.

EYE DISORDERS (GLAUCOMA)

The ocular health of individuals is seriously threatened by eye conditions, such as glaucoma, which, if unchecked, can result in irreparable vision loss. Damage to the optic nerve, a characteristic of the heterogeneous group of visual illnesses known as glaucoma, is frequently caused by elevated intraocular pressure. This increased pressure may harm the optic nerve, leading to progressive vision loss over time. To avoid or delay the start of visual loss and to ensure ideal ocular health, it is crucial to treat glaucoma effectively.

Disease Overview

Glaucoma is primarily caused by an imbalance between aqueous humor production and drainage, which results in an increase in intraocular pressure (IOP) [68]. A gradual loss of peripheral vision, tunnel vision, and eye pain are among the common symptoms of this disorder. In severe situations, it can even cause total vision loss [69]. Glaucoma has been identified as the most common cause of permanent blindness, with an estimated global impact of 79.6 million individuals [70]. It is essential to investigate this at the molecular level to comprehend the underlying causes of this disease. the α2-adrenergic receptors (α2-ARs), which are found in the ciliary body of the eye, serve as the primary targets for glaucoma. The regulation of aqueous humor production and drainage is dependent on the activation of these receptors, and their activation leads to a decrease in IOP [71].

The computer method known as the "molecular docking approach" is used to analyze protein-ligand interactions and forecast the binding affinity of small molecules to a target protein [72]. *In silico* docking involves generating a multitude of ligand conformations, evaluating their interaction with the target protein, and scoring the resulting complexes based on their predicted stability [73]. The primary objective of this process is to identify compounds that exhibit high binding affinities and optimize their structures to enhance their efficacy and reduce potential side effects [74].

Case Study: Brimonidine

Molecular docking methods were used to create the alpha-2 adrenergic receptor (α2-AR) agonist brimonidine to specifically target glaucoma [75]. By selectively binding to α2-ARs located in the ciliary body, this pharmaceutical agent effectively decreases the production of aqueous humor while simultaneously improving its drainage [71]. As a result of this mechanism, brimonidine has been successful in mitigating intraocular pressure (IOP), thus prohibiting the advancement of glaucoma.

The drug development process that incorporates molecular docking involves numerous stages, starting with target identification and then proceeding to virtual screening, hit identification, lead compound optimization, and preclinical testing [76]. Molecular docking techniques allow researchers to identify prospective lead compounds that possess high binding affinities for the target protein [77]. These leads then undergo further optimization and testing to improve their pharmacokinetic and pharmacodynamic properties before advancing to preclinical and clinical trials [78].

Clinical Trials and Approval

Brimonidine, a medication utilized for treating glaucoma, underwent a series of extensive and meticulous clinical trials to evaluate its safety and efficacy in the context of this ocular malady [79]. Such trials, conducted with scrupulous attention to detail and scientific rigor to ensure the veracity of their findings, ultimately demonstrated that brimonidine was capable of significantly reducing intraocular pressure (IOP) and exhibited a highly satisfactory safety profile. Brimonidine was approved by the US Food and Drug Administration (FDA) in 1996 thanks in large part to the abundance of empirical data supporting its safety and efficacy as a therapy for glaucoma [80].

Real-world Impact on Disease Treatment

Brimonidine has become a widely accepted method of treatment for the practical management of glaucoma and ocular hypertension, demonstrating a highly advantageous alternative to other medications that lower intraocular pressure (IOP) [81]. Its practical impact in the real world is exemplified by the conservation of eyesight and the amelioration of the standard of living for a substantial multitude of patients grappling with the aforementioned ophthalmological condition [82].

CHALLENGES AND LIMITATIONS IN THE USE OF MOLECULAR DOCKING FOR DRUG DEVELOPMENT AND STRATEGIES TO OVERCOME THEM

A highly useful method in the realm of drug discovery is molecular docking, which refers to the process of finding potential novel therapies and creating medications. Its utility lies in its ability to sift through vast databases of compounds to identify those with the highest likelihood of binding. Nonetheless, this technique is not without its challenges and limitations. One of the primary challenges associated with molecular docking is the issue of sampling bias and false positives. The random sampling method on which it is based may result in biased outcomes and inaccurate predictions. Contrary to the usual force field computation that is typically used in molecular docking, the consensus scoring function has been shown to greatly improve docking accuracy. Research in this field has been conducted to support this finding [83].

Due to its intricacy, the drug-target interaction poses a substantial challenge. Accurately predicting the binding affinity of a possible drug-target pair is challenging due to its complexity. Indeed, a number of elements, including shape, charge, hydrophobicity, and even hydrogen bonding, affect how the two molecules interact. Furthermore, the quantity of ligands and the type of target can both affect how complicated the drug-target combination is (*e.g.*, protein, DNA). In addition, the computational time and resources required for successful molecular docking should not be underestimated. Even with the reduction in time needed for a successful binding prediction due to increased computational resources, it remains a time-intensive process requiring significant resources. To overcome these challenges, it is critical to consider sophisticated algorithms and approaches. For instance, fast Fourier transform (FFT) can be employed in molecular docking to calculate the correlation between the shape of the ligand and the receptor. Additionally, shape-independent and energy-based algorithms, along with artificial intelligence (AI) and machine learning (ML) technology, have greatly enhanced a number of aspects, including treatment, medication, screening, prediction, forecasting, contact tracing, and the process of developing drugs and vaccines for the COVID-19 pandemic [84].

Through the integration of protein structural knowledge and pharmacophore modeling, docking simulation accuracy can be improved. These methods reveal important interaction locations and fundamental characteristics necessary for a ligand to successfully connect to its target protein, providing new information about the protein-ligand binding process. As a result, docking simulations can concentrate on pertinent locations and produce more precise predictions. An illustration of this is the Pharmer method, which has demonstrated effectiveness in drug development by quickly screening enormous compound libraries and selecting molecules with particular chemical characteristics essential for their biological activity as pharmacophores [85].

Furthermore, by developing more complex scoring systems and increasing the resolution of computational models, the accuracy and speed of computations can be improved. In docking tests against widely used docking programs such as Dock, FlexX, and Gold, some scoring functions have already shown superior performance [86].

CONCLUSION

The field of technology and computing capacities has advanced quickly in recent years, making it possible to use molecular docking as a productive and affordable method for the identification of novel medications and treatments. The viability of molecular docking as a method for virtual screening has been established. Furthermore, when used within the scope of their applicability, advanced algorithms and structure-based scoring functions for binding affinity prediction have been found to outperform conventional scoring functions. Additionally, the integration of artificial intelligence and machine learning with molecular docking might further improve prediction accuracy and make it possible to identify prospective medication candidates quickly. Overall, it is clear that molecular docking has enormous potential to accelerate the drug discovery process. Therefore, it is strongly advised that additional research be done in this area to fully realize its potential.

REFERENCES

[1]   Sousa SF, Fernandes PA, Ramos MJ. Protein–ligand docking: Current status and future challenges. Proteins 2006; 65(1): 15-26.
        [http://dx.doi.org/10.1002/prot.21082] [PMID: 16862531]

[2]   Gohlke H, Hendlich M, Klebe G. Knowledge-based scoring function to predict protein-ligand interactions. J Mol Biol 2000; 295(2): 337-56.
        [http://dx.doi.org/10.1006/jmbi.1999.3371] [PMID: 10623530]

[3]   Cava C, Castiglioni I. Integration of molecular docking and *in vitro* studies: A powerful approach for drug discovery in breast cancer. Appl Sci 2020; 10(19): 6981.
        [http://dx.doi.org/10.3390/app10196981]

[4]   Patel SB, Patel BD, Pannecouque C, Bhatt HG. Design, synthesis and anti-HIV activity of novel quinoxaline derivatives. Eur J Med Chem 2016; 117: 230-40.
        [http://dx.doi.org/10.1016/j.ejmech.2016.04.019] [PMID: 27105027]

[5]   Guleria V, Pal T, Sharma B, Chauhan S, Jaiswal V. Pharmacokinetic and molecular docking studies to design antimalarial compounds targeting Actin I. Int J Health Sci (Qassim) 2021; 15(6): 4-15.
        [PMID: 34916893]

[6]   Trott O, Olson AJ. AutoDock Vina: Improving the speed and accuracy of docking with a new scoring function, efficient optimization, and multithreading. J Comput Chem 2010; 31(2): 455-61.
        [http://dx.doi.org/10.1002/jcc.21334] [PMID: 19499576]

[7]   Hossain A, Rahman ME, Rahman MS, *et al.* Identification of medicinal plant-based phytochemicals as a potential inhibitor for SARS-CoV-2 main protease (M^{pro}) using molecular docking and deep learning methods. Comput Biol Med 2023; 157: 106785.
        [http://dx.doi.org/10.1016/j.compbiomed.2023.106785] [PMID: 36931201]

[8]   Salmaso V, Moro S. Bridging molecular docking to molecular dynamics in exploring ligand-protein recognition process: An overview. Front Pharmacol 2018; 9: 923.

[9]   Porter KA, Xia B, Beglov D, *et al.* ClusPro PeptiDock: Efficient global docking of peptide recognition motifs using FFT. Bioinformatics 2017; 33(20): 3299-301.
        [http://dx.doi.org/10.1093/bioinformatics/btx216] [PMID: 28430871]

[10]   Jones G, Willett P, Glen RC. Molecular recognition of receptor sites using a genetic algorithm with a description of desolvation. J Mol Biol 1995; 245(1): 43-53.
[http://dx.doi.org/10.1016/S0022-2836(95)80037-9] [PMID: 7823319]

[11]   Axenopoulos A, Daras P, Papadopoulos G, *et al.* A shape descriptor for fast complementarity matching in molecular docking. IEEE/ACM Trans Comput Biol Bioinforma 2011; 8: 1441-57.
[http://dx.doi.org/10.1109/TCBB.2011.72]

[12]   Rush TS III, Grant JA, Mosyak L, Nicholls A. A shape-based 3-D scaffold hopping method and its application to a bacterial protein-protein interaction. J Med Chem 2005; 48(5): 1489-95.
[http://dx.doi.org/10.1021/jm040163o] [PMID: 15743191]

[13]   Sousa SF, Ribeiro AJM, Coimbra JTS, *et al.* Protein-ligand docking in the new millennium: A retrospective of 10 years in the field. Curr Med Chem 2013; 20(18): 2296-314.
[http://dx.doi.org/10.2174/0929867311320180002] [PMID: 23531220]

[14]   Fox GJ, Menzies D. A Review of the Evidence for Using Bedaquiline (TMC207) to Treat Multi-Drug Resistant Tuberculosis. Infect Dis Ther 2013; 2(2): 123-44.
[http://dx.doi.org/10.1007/s40121-013-0009-3] [PMID: 25134476]

[15]   Morel PA, Ta'asan S, Morel BF, Kirschner DE, Flynn JL. New insights into mathematical modeling of the immune system. Immunol Res 2006; 36(1-3): 157-66.
[http://dx.doi.org/10.1385/IR:36:1:157] [PMID: 17337776]

[16]   Mase S, Chorba T, Parks S, *et al.* Bedaquiline for the treatment of multidrug-resistant tuberculosis in the United States. Clin Infect Dis 2020; 71(4): 1010-6.
[http://dx.doi.org/10.1093/cid/ciz914] [PMID: 31556947]

[17]   Kumar S, Mehra R, Sharma S, *et al.* Screening of antitubercular compound library identifies novel ATP synthase inhibitors of *Mycobacterium tuberculosis*. Tuberculosis 2018; 108: 56-63.
[http://dx.doi.org/10.1016/j.tube.2017.10.008] [PMID: 29523328]

[18]   Andries K, Verhasselt P, Guillemont J, *et al.* A diarylquinoline drug active on the ATP synthase of *Mycobacterium tuberculosis*. Science 2005; 307(5707): 223-7.
[http://dx.doi.org/10.1126/science.1106753] [PMID: 15591164]

[19]   Siegel RL, Miller KD, Fuchs HE, Jemal A. Cancer Statistics, 2021. CA Cancer J Clin 2021; 71(1): 7-33.
[http://dx.doi.org/10.3322/caac.21654] [PMID: 33433946]

[20]   Bray F, Ferlay J, Soerjomataram I, Siegel RL, Torre LA, Jemal A. Global cancer statistics 2018: GLOBOCAN estimates of incidence and mortality worldwide for 36 cancers in 185 countries. CA Cancer J Clin 2018; 68(6): 394-424.
[http://dx.doi.org/10.3322/caac.21492] [PMID: 30207593]

[21]   Escudier B, Eisen T, Stadler WM, *et al.* Sorafenib in advanced clear-cell renal-cell carcinoma. N Engl J Med 2007; 356(2): 125-34.
[http://dx.doi.org/10.1056/NEJMoa060655] [PMID: 17215530]

[22]   Wilhelm SM, Carter C, Tang L, *et al.* BAY 43-9006 exhibits broad spectrum oral antitumor activity and targets the RAF/MEK/ERK pathway and receptor tyrosine kinases involved in tumor progression and angiogenesis. Cancer Res 2004; 64(19): 7099-109.
[http://dx.doi.org/10.1158/0008-5472.CAN-04-1443] [PMID: 15466206]

[23]   Llovet JM, Ricci S, Mazzaferro V, *et al.* Sorafenib in advanced hepatocellular carcinoma. N Engl J Med 2008; 359(4): 378-90.
[http://dx.doi.org/10.1056/NEJMoa0708857] [PMID: 18650514]

[24]   Paramashivam SK, Elayaperumal K, Natarajan B, Ramamoorthy M, Balasubramanian S, Dhiraviam K. In silico pharmacokinetic and molecular docking studies of small molecules derived from Indigofera aspalathoides Vahl targeting receptor tyrosine kinases. Bioinformation 2015; 11(2): 73-84.
[http://dx.doi.org/10.6026/97320630011073] [PMID: 25848167]

[25]   Meng F. Molecular dynamics simulation of VEGFR2 with sorafenib and other urea-substituted aryloxy compounds. J Theoret Chem 2013; 2013: 1-7.
[http://dx.doi.org/10.1155/2013/739574]

[26]   Mozaffarian D, Benjamin EJ, Go AS, *et al.* Heart disease and stroke statistics: 2016 update. Circulation 2016; 133(4): e38-e360.
[http://dx.doi.org/10.1161/CIR.0000000000000350] [PMID: 26673558]

[27]   Ho SJ, Brighton TA. Ximelagatran, direct thrombin inhibitor, oral anticoagulants, thromboprophylaxis. Vasc Health Risk Manag 2006; 2(1): 49-58.
[http://dx.doi.org/10.2147/vhrm.2006.2.1.49] [PMID: 17319469]

[28]   Antonsson T, Bylund R, Eriksson U, *et al.* Effects of melagatran, a new low-molecular-weight thrombin inhibitor, on thrombin and fibrinolytic enzymes. Thromb Haemost 1998; 79(1): 110-8.
[http://dx.doi.org/10.1055/s-0037-1614245] [PMID: 9459334]

[29]   Loganathan C, Sakkiah S, Lee K, Kabilan S, Meganathan C. Pharmacophore design, virtual screening, molecular docking and optimization approaches to discover potent thrombin inhibitors. Comb Chem High Throughput Screen 2013; 16(9): 702-20.
[http://dx.doi.org/10.2174/13862073113169990007] [PMID: 23713461]

[30]   Connolly SJ, Ezekowitz MD, Yusuf S, *et al.* Dabigatran *versus* warfarin in patients with atrial fibrillation. N Engl J Med 2009; 361(12): 1139-51.
[http://dx.doi.org/10.1056/NEJMoa0905561] [PMID: 19717844]

[31]   Kvien TK. Epidemiology and burden of illness of rheumatoid arthritis. PharmacoEconomics 2004; 22(S1) (1): 1-12.
[http://dx.doi.org/10.2165/00019053-200422001-00002] [PMID: 15157000]

[32]   Almutairi K, Nossent J, Preen D, Keen H, Inderjeeth C. The global prevalence of rheumatoid arthritis: A meta-analysis based on a systematic review. Rheumatol Int 2021; 41(5): 863-77.
[http://dx.doi.org/10.1007/s00296-020-04731-0] [PMID: 33175207]

[33]   Klareskog L, Catrina AI, Paget S. Rheumatoid arthritis. Lancet 2009; 373(9664): 659-72.
[http://dx.doi.org/10.1016/S0140-6736(09)60008-8] [PMID: 19157532]

[34]   Ghoreschi K, Jesson MI, Li X, *et al.* Modulation of innate and adaptive immune responses by tofacitinib (CP-690,550). J Immunol 2011; 186(7): 4234-43.
[http://dx.doi.org/10.4049/jimmunol.1003668] [PMID: 21383241]

[35]   van Vollenhoven RF, Fleischmann R, Cohen S, *et al.* Tofacitinib or adalimumab *versus* placebo in rheumatoid arthritis. N Engl J Med 2012; 367(6): 508-19.
[http://dx.doi.org/10.1056/NEJMoa1112072] [PMID: 22873531]

[36]   Changelian PS, Flanagan ME, Ball DJ, *et al.* Prevention of organ allograft rejection by a specific Janus kinase 3 inhibitor. Science 2003; 302(5646): 875-8.
[http://dx.doi.org/10.1126/science.1087061] [PMID: 14593182]

[37]   Sanachai K, Mahalapbutr P, Choowongkomon K, Poo-arporn RP, Wolschann P, Rungrotmongkol T. Insights into the binding recognition and susceptibility of tofacitinib toward janus kinases. ACS Omega 2020; 5(1): 369-77.
[http://dx.doi.org/10.1021/acsomega.9b02800] [PMID: 31956784]

[38]   Fleischmann R, Kremer J, Cush J, *et al.* Placebo-controlled trial of tofacitinib monotherapy in rheumatoid arthritis. N Engl J Med 2012; 367(6): 495-507.
[http://dx.doi.org/10.1056/NEJMoa1109071] [PMID: 22873530]

[39]   Traynor K. FDA approves tofacitinib for rheumatoid arthritis. Am J Heal Pharm 2012; 69(24): 2120.
[PMID: 23230026]

[40]   Schwartz DM, Kanno Y, Villarino A, Ward M, Gadina M, O'Shea JJ. JAK inhibition as a therapeutic strategy for immune and inflammatory diseases. Nat Rev Drug Discov 2017; 16(12): 843-62.

[http://dx.doi.org/10.1038/nrd.2017.201] [PMID: 29104284]

[41]   Drucker DJ, Nauck MA. The incretin system: Glucagon-like peptide-1 receptor agonists and dipeptidyl peptidase-4 inhibitors in type 2 diabetes. Lancet 2006; 368(9548): 1696-705.
[http://dx.doi.org/10.1016/S0140-6736(06)69705-5] [PMID: 17098089]

[42]   Vyshnavi P, Narayana P, Venkatesh P. Review on diabetes Mellitus. J Innov Appl Pharm Sci 2022; 7: 24-7.

[43]   Raz I, Hanefeld M, Xu L, Caria C, Williams-Herman D, Khatami H. Efficacy and safety of the dipeptidyl peptidase-4 inhibitor sitagliptin as monotherapy in patients with type 2 diabetes mellitus. Diabetologia 2006; 49(11): 2564-71.
[http://dx.doi.org/10.1007/s00125-006-0416-z] [PMID: 17001471]

[44]   Chakraborty C, Hsu MJ, Agoramoorthy G. Understanding the molecular dynamics of type-2 diabetes drug target DPP-4 and its interaction with Sitagliptin and inhibitor Diprotin-A. Cell Biochem Biophys 2014; 70(2): 907-22.
[http://dx.doi.org/10.1007/s12013-014-9998-0] [PMID: 24809328]

[45]   Goldstein BJ, Feinglos MN, Lunceford JK, Johnson J, Williams-Herman DE. Effect of initial combination therapy with sitagliptin, a dipeptidyl peptidase-4 inhibitor, and metformin on glycemic control in patients with type 2 diabetes. Diabetes Care 2007; 30(8): 1979-87.
[http://dx.doi.org/10.2337/dc07-0627] [PMID: 17485570]

[46]   Wahid SS, Ottman K, Bohara J, *et al.* Adolescent perspectives on depression as a disease of loneliness: A qualitative study with youth and other stakeholders in urban Nepal. Child Adolesc Psychiatry Ment Health 2022; 16(1): 51.
[http://dx.doi.org/10.1186/s13034-022-00481-y] [PMID: 35739569]

[47]   Stahl SM. The psychopharmacology of energy and fatigue. J Clin Psychiatry 2002; 63(1): 7-8.
[http://dx.doi.org/10.4088/JCP.v63n0102] [PMID: 11838630]

[48]   Bhat S, Newman AH, Freissmuth M. How to rescue misfolded SERT, DAT and NET: targeting conformational intermediates with atypical inhibitors and partial releasers. Biochem Soc Trans 2019; 47(3): 861-74.
[http://dx.doi.org/10.1042/BST20180512] [PMID: 31064865]

[49]   Baldwin DS, Necking O, Schmidt SN, Ren H, Reines EH. Efficacy and safety of vortioxetine in treatment of patients with major depressive disorder and common co-morbid physical illness. J Affect Disord 2022; 311: 588-94.
[http://dx.doi.org/10.1016/j.jad.2022.05.098] [PMID: 35597471]

[50]   Montgomery SA, Nielsen RZ, Poulsen LH, Häggström L. A randomised, double-blind study in adults with major depressive disorder with an inadequate response to a single course of selective serotonin reuptake inhibitor or serotonin–noradrenaline reuptake inhibitor treatment switched to vortioxetine or agomelatine. Hum Psychopharmacol 2014; 29(5): 470-82.
[http://dx.doi.org/10.1002/hup.2424] [PMID: 25087600]

[51]   Nackenoff AG, Simmler LD, Baganz NL, Pehrson AL, Sánchez C, Blakely RD. Serotonin transporter-independent actions of the antidepressant vortioxetine as revealed using the SERT Met172 mouse. ACS Chem Neurosci 2017; 8(5): 1092-100.
[http://dx.doi.org/10.1021/acschemneuro.7b00038] [PMID: 28272863]

[52]   Andersen J, Ladefoged LK, Wang D, *et al.* Binding of the multimodal antidepressant drug vortioxetine to the human serotonin transporter. ACS Chem Neurosci 2015; 6(11): 1892-900.
[http://dx.doi.org/10.1021/acschemneuro.5b00225] [PMID: 26389667]

[53]   Gonda X, Sharma SR, Tarazi FI. Vortioxetine: A novel antidepressant for the treatment of major depressive disorder. Expert Opin Drug Discov 2019; 14(1): 81-9.
[http://dx.doi.org/10.1080/17460441.2019.1546691] [PMID: 30457395]

[54]   Sowa-Kućma M, Pańczyszyn-Trzewik P, Misztak P, *et al.* Vortioxetine: A review of the

pharmacology and clinical profile of the novel antidepressant. Pharmacol Rep 2017; 69(4): 595-601.
[http://dx.doi.org/10.1016/j.pharep.2017.01.030] [PMID: 28499187]

[55]   McKone EF, Aitken ML. Cystic fibrosis: Disease mechanisms and therapeutic targets. Drug Discov Today Dis Mech 2004; 1(1): 137-43.
[http://dx.doi.org/10.1016/j.ddmec.2004.08.012]

[56]   Righetti G, Casale M, Liessi N, *et al.* Molecular docking and QSAR studies as computational tools exploring the rescue ability of F508DEL CFTR correctors. Int J Mol Sci 2020; 21(21): 8084.
[http://dx.doi.org/10.3390/ijms21218084] [PMID: 33138251]

[57]   Molinski SV, Shahani VM, Subramanian AS, *et al.* Comprehensive mapping of cystic fibrosis mutations to CFTR protein identifies mutation clusters and molecular docking predicts corrector binding site. Proteins 2018; 86(8): 833-43.
[http://dx.doi.org/10.1002/prot.25496] [PMID: 29569753]

[58]   Mijnders M, Kleizen B, Braakman I. Correcting CFTR folding defects by small-molecule correctors to cure cystic fibrosis. Curr Opin Pharmacol 2017; 34: 83-90.
[http://dx.doi.org/10.1016/j.coph.2017.09.014] [PMID: 29055231]

[59]   Cholon DM, Quinney NL, Fulcher ML, *et al.* Potentiator ivacaftor abrogates pharmacological correction of ΔF508 CFTR in cystic fibrosis. Sci Transl Med 2014; 6(246): 246-96.
[http://dx.doi.org/10.1126/scitranslmed.3008680]

[60]   Bulloch MN, Hanna C, Giovane R. Lumacaftor/ivacaftor, a novel agent for the treatment of cystic fibrosis patients who are homozygous for the F580del CFTR mutation. Expert Rev Clin Pharmacol 2017; 10(10): 1055-72.
[http://dx.doi.org/10.1080/17512433.2017.1378094] [PMID: 28891346]

[61]   Wainwright CE, Elborn JS, Ramsey BW, *et al.* Lumacaftor–Ivacaftor in Patients with Cystic Fibrosis Homozygous for Phe508del *CFTR*. N Engl J Med 2015; 373(3): 220-31.
[http://dx.doi.org/10.1056/NEJMoa1409547] [PMID: 25981758]

[62]   Yasothan U, Kar S. Osteoporosis: overview and pipeline. Nat Rev Drug Discov 2008; 7(9): 725-6.
[http://dx.doi.org/10.1038/nrd2620] [PMID: 19172687]

[63]   Sandhu SK, Hampson G. The pathogenesis, diagnosis, investigation and management of osteoporosis. J Clin Pathol 2011; 64(12): 1042-50.
[http://dx.doi.org/10.1136/jcp.2010.077842] [PMID: 21896577]

[64]   Pouresmaeili F, Kamali Dehghan B, Kamarehei M, Yong Meng G. A comprehensive overview on osteoporosis and its risk factors. Ther Clin Risk Manag 2018; 14: 2029-49.
[http://dx.doi.org/10.2147/TCRM.S138000] [PMID: 30464484]

[65]   Drake FH, Dodds RA, James IE, *et al.* Cathepsin K, but not cathepsins B, L, or S, is abundantly expressed in human osteoclasts. J Biol Chem 1996; 271(21): 12511-6.
[http://dx.doi.org/10.1074/jbc.271.21.12511] [PMID: 8647859]

[66]   Gauthier JY, Chauret N, Cromlish W, *et al.* The discovery of odanacatib (MK-0822), a selective inhibitor of cathepsin K. Bioorg Med Chem Lett 2008; 18(3): 923-8.
[http://dx.doi.org/10.1016/j.bmcl.2007.12.047] [PMID: 18226527]

[67]   McClung MR, O'Donoghue ML, Papapoulos SE, *et al.* Odanacatib for the treatment of postmenopausal osteoporosis: Results of the LOFT multicentre, randomised, double-blind, placebo-controlled trial and LOFT Extension study. Lancet Diabetes Endocrinol 2019; 7(12): 899-911.
[http://dx.doi.org/10.1016/S2213-8587(19)30346-8] [PMID: 31676222]

[68]   Quigley HA. Glaucoma. Lancet 2011; 377(9774): 1367-77.
[http://dx.doi.org/10.1016/S0140-6736(10)61423-7] [PMID: 21453963]

[69]   Jonas JB, Aung T, Bourne RR, Bron AM, Ritch R, Panda-Jonas S. Glaucoma. Lancet 2017; 390(10108): 2183-93.
[http://dx.doi.org/10.1016/S0140-6736(17)31469-1] [PMID: 28577860]

[70]   Tham YC, Li X, Wong TY, Quigley HA, Aung T, Cheng CY. Global prevalence of glaucoma and projections of glaucoma burden through 2040: A systematic review and meta-analysis. Ophthalmol 2014; 121(11): 2081-90.
[http://dx.doi.org/10.1016/j.ophtha.2014.05.013] [PMID: 24974815]

[71]   Cantor LB. Brimonidine in the treatment of glaucoma and ocular hypertension. Ther Clin Risk Manag 2006; 2(4): 337-46.
[http://dx.doi.org/10.2147/tcrm.2006.2.4.337] [PMID: 18360646]

[72]   Meng XY, Zhang HX, Mezei M, Cui M. Molecular docking: a powerful approach for structure-based drug discovery. Curr Computeraided Drug Des 2011; 7(2): 146-57.
[http://dx.doi.org/10.2174/157340911795677602] [PMID: 21534921]

[73]   Kitchen DB, Decornez H, Furr JR, Bajorath J. Docking and scoring in virtual screening for drug discovery: Methods and applications. Nat Rev Drug Discov 2004; 3(11): 935-49.
[http://dx.doi.org/10.1038/nrd1549] [PMID: 15520816]

[74]   Yuriev E, Ramsland PA. Latest developments in molecular docking: 2010–2011 in review. J Mol Recognit 2013; 26(5): 215-39.
[http://dx.doi.org/10.1002/jmr.2266] [PMID: 23526775]

[75]   Giovannitti JA Jr, Thoms SM, Crawford JJ. Alpha-2 adrenergic receptor agonists: A review of current clinical applications. Anesth Prog 2015; 62(1): 31-8.
[http://dx.doi.org/10.2344/0003-3006-62.1.31] [PMID: 25849473]

[76]   Ferreira L, dos Santos R, Oliva G, Andricopulo A. Molecular docking and structure-based drug design strategies. Molecules 2015; 20(7): 13384-421.
[http://dx.doi.org/10.3390/molecules200713384] [PMID: 26205061]

[77]   Pagadala NS, Syed K, Tuszynski J. Software for molecular docking: A review. Biophys Rev 2017; 9(2): 91-102.
[http://dx.doi.org/10.1007/s12551-016-0247-1] [PMID: 28510083]

[78]   Lionta E, Spyrou G, Vassilatis D, Cournia Z. Structure-based virtual screening for drug discovery: principles, applications and recent advances. Curr Top Med Chem 2014; 14(16): 1923-38.
[http://dx.doi.org/10.2174/1568026614666140929124445] [PMID: 25262799]

[79]   Schuman JS, Horwitz B, Choplin NT, David R, Albracht D, Chen K. A 1-year study of brimonidine twice daily in glaucoma and ocular hypertension. A controlled, randomized, multicenter clinical trial. Arch Ophthalmol 1997; 115(7): 847-52.
[http://dx.doi.org/10.1001/archopht.1997.01100160017002] [PMID: 9230823]

[80]   Oh DJ, Chen JL, Vajaranant TS, Dikopf MS. Brimonidine tartrate for the treatment of glaucoma. Expert Opin Pharmacother 2019; 20(1): 115-22.
[http://dx.doi.org/10.1080/14656566.2018.1544241] [PMID: 30407890]

[81]   Schwartz GF, Quigley HA. Adherence and persistence with glaucoma therapy. Surv Ophthalmol 2008; 53(6) (1): S57-68.
[http://dx.doi.org/10.1016/j.survophthal.2008.08.002] [PMID: 19038625]

[82]   Jackson JM, Knuckles M, Minni J, Johnson S, Belasco K. The role of brimonidine tartrate gel in the treatment of rosacea. Clin Cosmet Investig Dermatol 2015; 8: 529-38.
[http://dx.doi.org/10.2147/CCID.S58920] [PMID: 26566370]

[83]   Wang R, Lai L, Wang S. Further development and validation of empirical scoring functions for structure-based binding affinity prediction. J Comput Aided Mol Des 2002; 16(1): 11-26.
[http://dx.doi.org/10.1023/A:1016357811882] [PMID: 12197663]

[84]   Lalmuanawma S, Hussain J, Chhakchhuak L. Applications of machine learning and artificial intelligence for Covid-19 (SARS-CoV-2) pandemic: A review. Chaos Solitons Fractals 2020; 139: 110059.
[http://dx.doi.org/10.1016/j.chaos.2020.110059] [PMID: 32834612]

[85]   Koes DR, Camacho CJ. Pharmer: Efficient and exact pharmacophore search. J Chem Inf Model 2011; 51(6): 1307-14.
[http://dx.doi.org/10.1021/ci200097m] [PMID: 21604800]

[86]   Muryshev AE, Tarasov DN, Butygin AV, Butygina OY, Aleksandrov AB, Nikitin SM. A novel scoring function for molecular docking. J Comput Aided Mol Des 2003; 17(9): 597-605.
[http://dx.doi.org/10.1023/B:JCAM.0000005766.95985.7e] [PMID: 14713191]

CHAPTER 3

The Roles of Farnesol and Farnesene in Curtailing Antibiotic Resistance

Axel R. Molina-Gallardo[1], Yesica R. Cruz-Martínez[1], Julieta Orozco-Martínez[1], Israel Valencia Quiroz[2] and C. Tzasna Hernández-Delgado[1,*]

[1] *Laboratory of Natural Products Bioactivity, UBIPRO, Superior Studies Faculty (FES)-Iztacala, National Autonomous University of México (UNAM), Tlalnepantla de Baz, México State, 54090, México*

[2] *Phytochemistry Laboratory, UBIPRO, Superior Studies Faculty (FES)-Iztacala, National Autonomous University of México (UNAM), Tlalnepantla de Baz, México State, 54090, México*

Abstract: In the extensive domain of "biotechnology and drug development for targeting human diseases", essential oils have long been revered for their therapeutic potential. Among these, farnesol and farnesene stand out due to their pharmacological attributes. As the challenge of antibiotic resistance intensifies, the scientific community is increasingly exploring the potential of these traditional remedies. Using the Kirby-Bauer agar diffusion method, a qualitative assessment was conducted on two gram-positive and two gram-negative bacterial strains. The broth microdilution technique further determined the Minimum Inhibitory Concentration (MIC), Minimum Bactericidal Concentration (MBC), and the sensitizing impacts of these compounds. Both farnesol and farnesene exhibited antibacterial efficacy against all evaluated strains. Their synergistic potential was highlighted when combined with clavulanic acid, cefuroxime, and cefepime. Among these combinations, farnesene paired with cefepime showed pronounced efficacy against *Escherichia coli* 82 MR, with an MIC of 0.47 µg/mL. In contrast, in the investigation of *Staphylococcus aureus* 23MR, it was observed that this particular strain exhibited an increased sensitivity when exposed to combinations containing farnesol. Notably, the Minimum Inhibitory Concentration (MIC) was determined to be 0.03 µg/mL in the presence of both antibiotic agents. To gain deeper molecular insights, docking experiments were performed with the *β*-lactamases of *E. coli* and *S. aureus*, focusing on the most effective combinations. All tested compounds—cefuroxime, cefepime, farnesene, and farnesol—acted as non-competitive inhibitors, suggesting their potential mechanisms of action.

Keywords: Antibacterial, Antibiofilm, Essential oils, Farnesol, Farnesene.

* **Corresponding author C. Tzasna Hernández-Delgado:** Laboratory of Natural Products Bioactivity, UBIPRO, Superior Studies Faculty (FES)-Iztacala, National Autonomous University of México (UNAM), Tlalnepantla de Baz, México State, 54090, México; Tel: +525527288013; E-mail: tzasna@unam.mx

INTRODUCTION

In the realm of microbiology, the emergence of antibiotic resistance serves as a compelling illustration of the remarkable adaptability and tenacity exhibited by bacterial organisms. As they naturally evolve, bacteria undergo mutations, acquire genetic material, and modify their genome expression. Such evolutionary changes empower them to counteract the effects of antibiotics [1]. While this evolutionary process is a natural occurrence, human activities have accelerated its pace. Within the broader context of microbial evolution, several determinants have been identified as contributors to the proliferation of antimicrobial-resistant bacterial strains. These encompass the imprudent utilization of medications in both human and veterinary medicine, suboptimal infection prevention protocols in healthcare establishments, and lapses in fundamental hygiene and sanitation standards. This challenge is further intensified by the declining discovery of new antibiotics [2].

Due to this situation, there has been a significant rise in the search for complementary treatments or alternatives to conventional antibiotics. Natural antimicrobial products have emerged as substances of significant scientific interest due to their vast chemical diversity and biological characteristics. Among these, essential oils include a group of components with unique qualities that distinguish them in the healthcare, cosmetics, and food industries [3].

In the broader discourse on essential oils featured in botanical literature, the sesquiterpene alcohol, farnesol ($C_{15}H_{26}O$), emerges as a notable compound. It has been cataloged in a range of plant families, prominently among Lamiaceae [4, 5], Asteraceae, Poaceae [6], and Pittosporaceae [7], among others [8]. In a parallel vein, farnesene ($C_{15}H_{24}$), another representative of the sesquiterpene class, is traced back to its origin in the mevalonic acid biosynthetic pathway. Essential oils from various families like Lamiaceae [4, 5], Lauraceae [9], Meliaceae [10], Asteraceae [6, 11], Verbenaceae [12], and Euphorbiaceae [13] have been found to contain several isomers of farnesene. These isomers encompass a range of properties, including antimicrobial activities, attractant capabilities, signaling functions, and hormonal attributes.

Terpenes, the broader category to which both farnesol and farnesene belong, possess a plethora of biological properties. Their potential therapeutic applications span from antifungal, antibacterial, antiviral, antitumor, antiparasitic, hypoglycemic, anti-inflammatory, to analgesic effects. Such a diverse range of properties underscore the increasing interest in terpenes within the scientific community [14].

Recent studies have shed light on the antimicrobial potential of farnesol. For instance, its efficacy against strains of *Paracoccidioides brasiliensis* was evaluated, revealing its ability to induce cytoplasmic degeneration and inhibit germ tube formation [7]. Furthermore, the combined use of *tt*-farnesol and minocycline demonstrated a significant inhibitory effect on the growth of *S. aureus* ATCC43300 [15].

The potential of farnesol to influence antibiotic activity has been a topic of academic inquiry. Notably, in experiments where farnesol was paired with antibiotics such as ampicillin and oxacillin, evidence emerged of its capacity to augment the effectiveness of these drugs against methicillin-resistant strains of *S. aureus* [16]. Such findings suggest that farnesol could play a pivotal role in overcoming antibiotic resistance, especially in bacterial strains like methicillin-resistant *S. aureus* [15].

MATERIALS AND METHODS

In the expansive domain of biotechnology and drug development, the quest to unearth natural compounds and their potential synergy against bacterial strains is a pressing endeavor. This chapter embarks on an investigative journey into two such intriguing compounds, farnesol and farnesene, and their prospective roles in the battle against antibiotic resistance.

Our narrative commenced with the foundational elements. The compounds under investigation, including farnesol (CAS 4602-84-0), farnesene 90% (CAS W383902), and several other chemical agents, were procured from Sigma–Aldrich, a reputable supplier in Toluca, Mexico. The bacterial protagonists for this study included strains of *Escherichia coli* ATCC25922, *Staphylococcus aureus* ATCC29213, *E. coli* 82 MR, and *S. aureus* 23 MR multi-resistant strains. These strains were diligently handled, maintained at 4°C in Mueller Hinton agar, and primed for subsequent experiments.

The first experiment conducted was the antibacterial activity assay. The Petri dish is an environment where bacteria interact and respond to various compounds. The disc diffusion method, a time-honored technique in microbiology, was employed to determine the response of our bacterial strains to the presence of farnesol, farnesene, and other agents [17, 18]. Clear zones of inhibition, representing areas where bacterial growth was notably suppressed, serve as an indicator of the antibacterial efficacy of the compounds.

However, we had additional experiments to perform. To quantify the efficacy of these compounds, the broth dilution method was used [19, 20]. This technique facilitated the determination of the minimum inhibitory concentration (MIC) and minimum bactericidal concentration (MBC) – in essence, the minimal quantities of the compounds needed to inhibit or stop bacterial growth.

We then investigated the activity of farnesol and farnesene in preventing biofilm formation. Biofilms, structured communities of bacteria that adhere to surfaces and each other, are infamous for their resilience against antibiotics. The Gómez-Sequeda approach was used to achieve the anti-biofilm effect, and the results from this section hint at the potential of these compounds to dismantle these bacterial communities [21].

Moreover, we investigated the synergistic dynamics of farnesol and farnesene when combined with antibiotics. Could these natural entities amplify the potency of established antibiotics? The modified broth microdilution technique was used to evaluate synergy, and the results unveiled captivating interactions that might provide innovative therapeutic strategies [22].

The bacterial growth trajectory, which is a depiction of the bacterial lifecycle, was also influenced by the presence of our compounds. The time-kill assay illuminated how farnesol and farnesene, either alone or in combination with antibiotics, influenced bacterial proliferation over a temporal spectrum [23].

Rigorous statistical evaluations fortified the credibility of the findings, thereby adding a layer of assurance to the narrative [24]. In addition, this work included molecular docking, a modality to elucidate how molecules, such as farnesol and farnesene, engage with proteins that are pivotal in antibiotic resistance. This molecular investigation, visualized *via* specialized software, provided insights into the potential mechanisms underlying the observed antibacterial effects [25].

In the overarching narrative of this study, this section accentuates the importance of probing natural compounds as potential allies in our fight against antibiotic resistance. As the global community confronts the waning efficacy of antibiotics, the revelations from this chapter provide hope by suggesting that nature might harbor the solutions to this therapeutic paradigm.

RESULTS

Antibacterial Activity in Qualitative Terms

In the intricate tapestry of biotechnology and drug development, the results section often serves as the heart of the narrative, revealing the potency and potential of the compounds under investigation. In this chapter, we delve into the antibacterial prowess of farnesol and farnesene, two natural compounds that have sparked scientific curiosity.

Tables **1** and **2** present the qualitative antibacterial activity of various compounds and combinations. At the core of this exploration were the following four bacterial strains: *S. aureus* ATCC 29213, *S. aureus* 23 MR, *E. coli* ATCC 25922, and *E. coli* 82 MR. These strains served as subjects for testing the antibacterial properties of farnesene (NP1) and farnesol (NP2) and their interactions with several antibiotics.

Table 1. Qualitative antibacterial activity of different compounds.

Strain	Compounds						
	Farnesene (NP1)	Farnesol (NP2)	Positive control (chloramphenicol)	Cefepime	Cefuroxime	Clavulanic Acid	Reserpine
S. aureus ATCC 29213	7±0.00	6±0.00	22±0.00	20±0.00	NA	6±0.00	NA
S. aureus MR 23	7±0.00	6±0.00	22±0.00	NA	NA	NA	NA
E. coli ATCC 25922	6±0.00	7±0.00	22±0.00	NA	NA	NA	NA
E. coli MR 82	6±0.00	6±0.00	22±0.00	NA	NA	6±0.00	NA

Inhibition zones (mm). NP= natural product. Mean ± standard deviation are reported. Concentrations (per disc): Control (+) = chloramphenicol 25 µg, cefuroxime 15 µg, cefepime 15 µg, reserpine 15 µg, clavulanic acid 15 µg, farnesol and farnesene 4 µL. NA= No activity.

Table 2. Qualitative antibacterial activity of different combinations with NPs.

Strain	Combinations with Farnesene				Combinations with Farnesol			
	Cefepime + NP1	Cefuroxime + NP1	Clavulanic Acid + NP1	Reserpine + NP1	Cefepime + NP2	Cefuroxime + NP2	Clavulanic Acid + NP2	Reserpine + NP2
S. aureus ATCC 29213	20±0.00	22±0.00	7±0.00	NA	6±0.00	20±0.00	6±0.00	NA
S. aureus MR 23	10±0.00	6±0.00	8±0.00	NA	7±0.00	9±0.00	6±0.00	NA
E. coli ATCC 25922	7±0.00	6±0.00	6±0.00	NA	7±0.00	6±0.00	6±0.00	NA

(Table 2) cont.....

Strain	Combinations with Farnesene				Combinations with Farnesol			
	Cefepime + NP1	Cefuroxime + NP1	Clavulanic Acid + NP1	Reserpine + NP1	Cefepime + NP2	Cefuroxime + NP2	Clavulanic Acid + NP2	Reserpine + NP2
E. coli MR 82	10±0.00	9±0.00	9±0.00	NA	6±0.00	6±0.00	6±0.00	NA

Inhibition zones (mm). NP= natural product. Mean ± standard deviation are reported. Concentrations (per disc): Control (+) = chloramphenicol 25 µg, cefuroxime 15 µg, cefepime 15 µg, reserpine 15 µg, clavulanic acid 15 µg, farnesol and farnesene 4 µL. NA= No activity.

The results paint a vivid picture. Farnesene (NP1) and farnesol (NP2) were examined both alone and in combination with several antibiotics. Chloramphenicol, a broad-spectrum antibiotic, served as the positive control in this study. The antibiotics tested were cefepime, cefuroxime, as well as references to clavulanic acid, and reserpine. While chloramphenicol consistently showed robust antibacterial activity, the inhibition zones for NP1 and NP2 were more modest but still indicative of their antibacterial potential. According to the data, inhibition zones for NP1 and NP2 ranged from 6.0 to 7.0 mm. However, compared to chloramphenicol, which consistently showed 22.0 mm inhibition zones against all four bacterial strains, their activities were noticeably lower.

Furthermore, when these natural compounds were paired with certain antibiotics, their combined antibacterial activity substantially increased or remained consistent. For instance, cefepime, cefuroxime, and clavulanic acid all displayed greater or comparable inhibitory zones when coupled with farnesene (NP1) as opposed to their independent activity. However, there was no activity seen when NP1 and reserpine were combined. The inhibition zones for cefepime and cefuroxime rose or remained the same when farnesol (NP2) was added to the antibiotics, while clavulanic acid activity remained unaffected. Like NP1, NP2, and reserpine did not exhibit any action.

It can be observed that in most cases, pure antibiotics did not show an inhibition zone, except for cefepime with a 20 mm inhibition zone in *S. aureus* ATCC 29213. However, farnesol and farnesene did present inhibition of growth of all strains. On the other hand, in all combinations of antibiotics and farnesol or farnesene, there is an activity in all bacterial strains except combinations with reserpine.

Antibacterial Activity in Quantitative Terms

Table **3** delves deeper into the results to quantify the antibacterial activity of farnesene and farnesol against four bacterial strains: *S. aureus* ATCC 29213, *S.*

aureus 23 MR, *E. coli* ATCC 25922, and *E. coli* 82MR 82 [20]. The metrics of interest here were the MIC and MBC. These values essentially gauge the smallest amount of the compounds needed to inhibit or stop bacterial growth.

Table 3. Minimum Inhibitory Concentration (MIC) and Minimum Bactericidal Concentration (MBC) (μg/mL) of farnesol and farnesene.

Strain	Farnesene		Farnesol	
	MIC	MBC	MIC	MBC
S. aureus ATCC 29213	1500	4000	>5000	>>5000
S. aureus MR 23	1500	5000	>5000	>>5000
E. coli ATCC 25922	1500	5000	>5000	>>5000
E. coli MR 82	1500	5000	>5000	>>5000

All strains examined had farnesene MICs of 1500 μg/mL. Almost all of them had MBCs of 5000 μg/mL, with the exception of the *S. aureus* strain ATCC 29213, which had an MBC of 4000 μg/mL. Similarly, all strains had the same farnesol MICs and MBCs of 5000 μg/mL.

The data revealed that while farnesene exhibited some antibacterial activity, the concentrations needed were relatively high. On the other hand, farnesol required even higher concentrations for its antibacterial effects, suggesting that it might be less potent than farnesene [21]. In subsequent experiments, to avoid doing MIC and MBC determinations at even higher doses, farnesol concentrations of 5000 μg/mL and 7000 μg/mL, corresponding to its MIC and MBC values, were utilized. This choice was taken because the relatively high farnesol concentrations needed to show antibacterial action would not be as useful in practical applications.

In conclusion, both farnesene and farnesol have demonstrated antibacterial action. However, farnesene appears to have a higher potency than farnesol. Their practical utility as antibacterial agents may be constrained by the relatively large concentrations needed to both inhibit and kill the bacterium.

Antibiofilm Effects of Farnesol and Farnesene

Biofilms, which are structured bacterial communities, are notoriously resilient against antibiotics, making the ability of a compound to inhibit biofilm formation crucial. Once the MICs and MBCs of farnesol and farnesene were determined, their antibiofilm activity was evaluated [22]. The results showed that farnesene, at sub inhibitory concentrations (½MIC), inhibited biofilm formation by 44.1% in gram-positive bacteria for the catalog strains and 31.92% for the resistant strains. When evaluating the MBC, the inhibition percentages rose to 68.86% and 58.5% respectively. For gram-negative bacteria, the inhibition percentages at ½MIC were 19.02% for the catalog and 24.2% for the resistant strain, and at MBC, they were 59.36% and 45.35% respectively.

In contrast, farnesol at ½MIC inhibited biofilm formation by 23.69% in gram-positive bacteria for the catalog strain and 25.61% for the resistant strain. Upon evaluating the MBC, the inhibition percentages were 32.03% and 35.76% respectively. For gram-negative bacteria, the inhibition percentages at ½MIC were 20.55% for the catalog strain and 17.6% for the resistant strain, and at MBC, they were 43.4% and 31.86% respectively [23].

While farnesene demonstrated significant biofilm inhibition, surpassing 50% in most strains, farnesol exhibited more modest rates, with none exceeding 43.4% for any strain. In the broader context of this book, these findings underscore the potential of farnesene and farnesol in the fight against antibiotic resistance. Their efficacy and the concentrations needed for optimal results vary, offering valuable perspectives on the potential roles of natural compounds in therapeutic strategies [24, 25].

The Sensitizing Effect of Farnesol and Farnesene

In the discussion of sensitizing effects, it is elucidated how certain compounds increase the vulnerability of bacteria to antibiotics. Tables **4** and **5** present an in-depth analysis of this phenomenon.

Table 4. MICs (µg/mL) of antibiotics used alone and when combined with farnesene at a dose of ½MIC.

Treatment	Strains			
	S. aureus ATCC 29213	*S. aureus* MR 23	*E. coli* ATCC 25922	*E. coli* MR 82
Cefuroxime	-	-	-	-
MIC antibiotic	15	15	120	120
MIC of the NP1	1500	1500	1500	1500
MIC of the combination	15	15	30	120
Cefepime	-	-	-	-
MIC antibiotic	0.9	0.9	0.9	0.9
MIC of the NP1	1500	1500	1500	1500
MIC of the combination	0.9	0.9	0.9	0.47

Table 5. MICs (in g/mL) of antibiotics alone and when combined with farnesol at ½MIC.

Treatment	Strains			
	S. aureus ATCC 29213	*S. aureus* MR 23	*E. coli* ATCC 25922	*E. coli* MR 82
Cefuroxime	-	-	-	-
MIC antibiotic	15	15	120	120
MIC of the NP2	>5000	>5000	>5000	>5000
MIC of the combination	0.9	0.03	3.75	3.75
Cefepime	-	-	-	-
MIC antibiotic	0.9	0.9	0.9	0.9
MIC of the NP2	>5000	>5000	>5000	>5000
MIC of the combination	0.9	0.03	0.9	0.06

For *E. coli* ATCC 25922, the combination of cefuroxime and farnesene led to a significant reduction in MIC, suggesting that when paired, they might be more potent than when used individually (Table **4**). Similarly, for the resistant *E. coli* strain, the combined effect was even more pronounced, hinting at the potential of these combinations in tackling resistant strains.

In the context of farnesol, the results were even more compelling. The combination of farnesol and cefuroxime resulted in a substantial decrease in the MICs against the resistant *S. aureus* strain (Table **5**). This indicates that farnesol might possess a unique capability to amplify the efficacy of certain antibiotics, especially against formidable strains like methicillin-resistant *S. aureus*.

Furthermore, the MIC values for both *S. aureus* strains (ATCC 29213 and 23MR) remained consistent at 15 µg/mL when cefuroxime and farnesene were combined at ½MIC. However, for *E. coli* ATCC 25922, the MIC value plummeted from 120 to 30 µg/mL, suggesting a synergistic effect between the two substances. When paired with farnesene, the MIC value for *E. coli* 82MR remained unchanged at 120 µg/mL.

The MIC values for *S. aureus* ATCC 29213, *S. aureus* 23MR, and *E. coli* ATCC 25922 for cefepime remained consistent at 0.9 µg/mL when farnesene was added at ½MIC. Interestingly, when cefepime and farnesene were combined, the MIC value for *E. coli* 82MR decreased from 0.9 to 0.47 µg/mL, suggesting an enhanced effect against this specific strain.

In conclusion, Tables **4** and **5** illustrate the differential responses of various bacterial strains when farnesene and farnesol are combined with cefuroxime and cefepime. The data suggests that the combination affects some strains more significantly than others, providing valuable insights into future therapeutic strategies.

Bacterial Growth Curves: A Deep Dive

Fig. (**1a**) through (**1d**) offer a visual exploration of the effects of farnesol, farnesene, and their combinations with antibiotics on bacterial growth. But what insights do these curves truly provide?

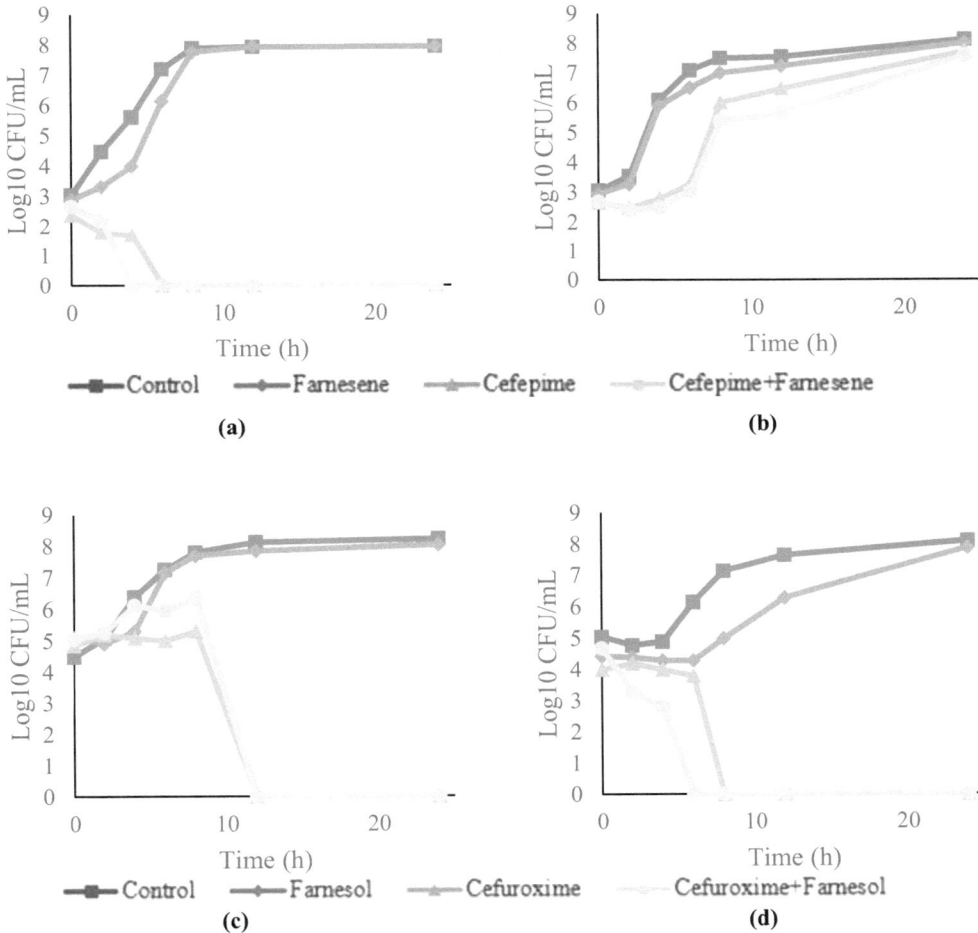

Fig. (1). Effects of farnesene on the growth curves of *E. coli* ATCC 25922 (**a**) and *E. coli* 82 MR (**b**), effects of farnesol on the growth curves of *S. aureus* ATCC 29213 (**c**) and *S. aureus* 23 MR (**d**). The concentrations used were: 1500 µg/mL (Farnesene), 0.9 µg/mL (Cefepime), and 0.9 µg/mL of cefepime + 750 µg/mL of farnesene; 5000 µg/mL (Farnesol), 15 µg/mL (Cefuroxime); and 15 µg/mL of cefuroxime + 2500 µg/mL of farnesol; the control contained no compounds. The number of survivors was calculated using the CFU count.

For the *E. coli* strain ATCC 25922 (Fig. **1a**), the combination of cefepime and farnesene exhibited a bactericidal effect, effectively eliminating the bacterial population. Interestingly, this combination acted faster, eradicating the bacteria at 4 hours, compared to cefepime alone, which took 6 hours. On its own, farnesene slowed the bacterial growth at intervals of 2, 4, and 6 hours, but after 8 hours, the growth resumed its usual pace.

When observing the *E. coli* 82MR strain (Fig. **1b**), the dynamics shifted. Both cefepime and its combination with farnesene initially displayed a bacteriostatic effect, pausing bacterial growth for the first 6 hours. Subsequent to this interval, the growth trajectory paralleled that of the control, though with a diminished population density. When utilized independently, farnesene exhibited a growth pattern comparable to the control, albeit with a reduced bacterial count.

In the examination of the *S. aureus* strain ATCC 29213, as depicted in Fig. (**1c**), both cefuroxime and its combination with farnesol exhibited bactericidal properties, eradicating the bacterial population by the 12-hour interval. However, the population density was discernibly reduced with cefuroxime in isolation compared to its combination with farnesol. Farnesol, in isolation, kept the growth consistent with the control but at a diminished population density.

Lastly, for the *S. aureus* 23MR strain (Fig. **1d**), the combination of cefuroxime and farnesol showcased bactericidal properties, exterminating the bacterial population at 6 hours, while cefuroxime alone achieved this by 8 hours. Farnesol, on its own, exhibited a bacteriostatic effect until the 6-hour point, after which the bacterial population surged.

In essence, these growth curves underscore the potential synergy between natural compounds and antibiotics, suggesting enhanced treatment efficacy for specific bacterial strains.

Molecular Docking: A Glimpse into the Molecular World

In the context of molecular interactions, molecular docking, visualized using the AutoDock 4.2 software [25], can be analogized to the precise fitting of a key into its corresponding lock. This technique offers an intricate comprehension of molecular interactions at the atomic scale. The results, summarized in Table **6** and visually represented in Fig. (**2a-d**), offer a comprehensive view of the interactions between these compounds and bacterial enzymes.

Table 6. Molecular affinity parameters of ligands with β-lactamases of *S. aureus* and *E. coli*.

Complex (protein-ligand)	Affinity (kcal/mol)	Ligand efficiency	Amino acids that interact through H+ bridges	Amino acids that interact through hydrophobic interactions
S. aureus β-lactamase (1BLC)				
1BLC/Farnesol	-4.88	-0.30	Ser176	Lys64, Arg65, Tyr172, Pro174, Tyr241 & Asn266
1BLC/Cefuroxime	-8.11	-0.28	Lys215, Leu212, Lys222, Lys227 & Val231	Lys222 & Val225
E. coli β-lactamase (3BLS)				
3BLS/Farnesene	-7.82	-0.52	-	Ile78, Lys84, Leu85, Leu254, Tyr259, Pro303, Pro304, & Thr305
3BLS/Cefepime	-7.01	-0.22	Gln98, His205 & Gln257	Phe97

Fig. (2). 3D representation of interactions between *S. aureus* β-lactamase residues and farnesol (**a**), *S. aureus* β-lactamase residues and cefuroxime (**b**), *E. coli* β-lactamase residues and farnesene (**c**) and *E. coli* β-lactamase residues and cefepime (**d**). Interactions are depicted as follows: blue lines (hydrogen bonds) and black lines (hydrophobic interactions). The illustrations were made using the Protein-Ligand Interaction Profiler software [26].

Regarding the *E. coli* strain ATCC 25922, the 1BLC/Farnesol complex displayed a notable affinity, characterized by a binding energy of -4.88 kcal/mol and a ligand efficiency of -0.30 as detailed in Table **6**. Fig. (**2a**) indicates that farnesol may act as a non-competitive inhibitor. This indicates that although it might not directly interact with the enzyme's active site, there exists a potential for it to impede the functionality of the enzyme. Similarly, the 1BLC/Cefuroxime complex displayed a binding energy of -8.11 kcal/mol and a ligand efficiency of -0.28 (Table **6**), hinting at its potential as a non-competitive inhibitor as visualized in Fig. (**2b**).

Upon further examination, the 3BLS/Farnesene complex demonstrated an affinity characterized by a binding energy of -7.82 kcal/mol and a ligand efficiency of -0.52, as detailed in Table **6**. Fig. (**2c**) suggests that farnesene, analogous to farnesol, has the potential to function as a non-competitive inhibitor. Lastly, the 3BLS/Cefepime complex, with a binding energy of -7.01 kcal/mol and a ligand efficiency of -0.22 (Table **6**), as depicted in Fig. (**2d**), further supports the notion of these compounds functioning as non-competitive inhibitors.

In essence, these molecular docking results underscore the potential synergy between natural compounds and antibiotics, suggesting enhanced treatment efficacy for specific bacterial strains.

DISCUSSION

In the realm of "biotechnology and drug development for targeting human diseases", the interplay between plants and their environment has given rise to a plethora of chemical compounds, each serving a distinct function. Among these, farnesol and farnesene stand out as components of essential oils found in various plants, serving multiple biological roles [4, 9, 12]. While their signaling roles have been previously acknowledged, this chapter delves deeper into their potential as antibacterial agents, specifically targeting *S. aureus* and *E. coli* strains [27].

The Role of Farnesene

The antimicrobial potential of farnesene is not just a hypothesis. Its presence in the coatings of fruits, especially apples, where it gives the characteristic green apple aroma, suggests its role [28]. However, when oxidized, it can produce compounds detrimental to the fruit, causing the outermost cell layers to die—a phenomenon termed scorching. Interestingly, this effect is also observed in prokaryotic cells, suggesting the potential of farnesene as an antimicrobial agent.

Farnesol's Antibacterial Potential

The ability of farnesol to inhibit the growth of microorganisms, particularly *S. aureus* and *E. coli*, has been well- documented [7, 29, 30]. However, its efficacy varies across pathogens. For instance, bacteria such as *Streptococcus mutans* and *Candida dubliniensis* exhibit higher tolerance [31, 32]. A significant mechanism of farnesol action involves its capacity to infiltrate biofilms, accumulate within cell membranes, and enhance membrane porosity [33]. This increase in permeability can enhance antibiotic uptake, a hypothesis supported by our findings. Furthermore, molecular docking experiments indicate that farnesol might interact with the β-lactamase enzyme, thereby further enhancing its antibacterial potential.

Mechanisms of Resistance and Sensitization

The debate around farnesol's role as an efflux pump inhibitor continues [34]. However, its effect when paired with clavulanic acid—a recognized β-lactamase inhibitor—hints that resistance mechanisms in specific strains might be related to β-lactamases [35]. The hydrophobic nature of farnesol allows for it to accumulate in the membrane, leading to leakage and potential biofilm destabilization [33, 36 - 38]. This property is further emphasized by its ability to inhibit biofilm formation in various strains, and farnesene shows even greater antibiofilm efficacy.

Synergistic Effects of Antibiotics

The combination of farnesol with antibiotics such as minocycline has shown promising results in previous studies [15]. In this investigation, the combination of farnesol with cefuroxime led to a significant reduction in the MICs against the resistant *S. aureus* strain. Similarly, farnesene led to enhanced antibiotic potency against *E. coli* strains, especially with cefepime.

Molecular Insights

Molecular docking results provided a deeper understanding of the interactions between these natural compounds and bacterial enzymes (Table **6**). These data suggest that farnesol might employ a non-competitive inhibitory mechanism against *S. aureus* β-lactamase (1BLC), allowing for antibiotics such as cefuroxime to exert their bacteriolytic activity. Meanwhile, farnesene might act as a competitive inhibitor against *E. coli* β-lactamase (3BLS), thereby enhancing the activity of antibiotics such as cefepime.

CONCLUSION

Within the framework of "biotechnology and drug development for targeting human diseases", the exploration of naturally derived compounds has provided valuable insights into potential therapeutic agents. Among these, farnesene has been identified as a compound of significant interest due to its pronounced antibacterial activity, as evidenced by its notable MIC values and inhibition percentages. Its efficacy in addressing bacterial challenges and inhibiting biofilm formation underscores its potential in the realm of therapeutic applications.

In contrast, farnesol, while not as potent in direct antibacterial interventions, has demonstrated a unique capability as a resistance modulator, particularly against resistant strains such as resistant *S. aureus*. When combined with specific antibiotics, such as cefepime, farnesol exhibits a synergistic effect, leading to substantial reductions in MIC values, highlighting its potential to enhance the efficacy of existing antibiotics.

From a molecular perspective, both farnesene and farnesol, especially when paired with antibiotics like cefuroxime, function as non-competitive inhibitors. Their targeted action against the β-lactamase resistance mechanism, a well-documented resistance pathway in several pathogens, offers promising avenues for future research and applications.

Given the increasing prevalence of nosocomial infections and the emergence of antibiotic-resistant strains, the need for innovative therapeutic solutions is paramount. In this context, the chapter has emphasized the potential contributions of farnesol and farnesene. As compounds derived from nature, they present promising avenues for the development of novel therapeutic strategies, potentially playing a pivotal role in our collective efforts to address the challenges of antibiotic resistance.

REFERENCES

[1]   Sultan I, Rahman S, Jan AT, Siddiqui MT, Mondal AH, Haq QMR. Antibiotics, resistome and resistance mechanisms: A bacterial perspective. Front Microbiol 2018; 9: 2066.
        [http://dx.doi.org/10.3389/fmicb.2018.02066] [PMID: 30298054]

[2]   Hiltunen T, Virta M, Laine AL. Antibiotic resistance in the wild: An eco-evolutionary perspective. Philos Trans R Soc Lond B Biol Sci 2017; 372(20160039): 2066.

[3]   Roohinejad S, Koubaa M, Barba F, *et al.* Extraction methods of essential oils from herbs and spices. Essential Oils in Food Processing Chemistry: Safety and Applications. 2017. USA: John Wiley Sons Ltd 2017; pp. 21-55.
        [http://dx.doi.org/10.1002/9781119149392.ch2]

[4]   Bozin B, Mimica-Dukic N, Simin N, Anackov G. Characterization of the volatile composition of essential oils of some lamiaceae spices and the antimicrobial and antioxidant activities of the entire oils. J Agric Food Chem 2006; 54(5): 1822-8.

[http://dx.doi.org/10.1021/jf051922u] [PMID: 16506839]

[5]   Uritu CM, Mihai CT, Stanciu GD, *et al.* Medicinal plants of the family lamiaceae in pain therapy: A review. Pain Res Manag 2018; 2018: 1-44.
[http://dx.doi.org/10.1155/2018/7801543] [PMID: 29854039]

[6]   Duarte MCT, Figueira GM, Sartoratto A, Rehder VLG, Delarmelina C. Anti-*Candida* activity of Brazilian medicinal plants. J Ethnopharmacol 2005; 97(2): 305-11.
[http://dx.doi.org/10.1016/j.jep.2004.11.016] [PMID: 15707770]

[7]   Derengowski LS, De-Souza-Silva C, Braz SV, *et al.* Antimicrobial effect of farnesol, a *Candida albicans* quorum sensing molecule, on *Paracoccidioides brasiliensis* growth and morphogenesis. Ann Clin Microbiol Antimicrob 2009; 8(1): 13.
[http://dx.doi.org/10.1186/1476-0711-8-13] [PMID: 19402910]

[8]   Eslahi H, Fahimi N, In A. Extraction methods of essential oils from herbs and spices. 2017. Bagher SMB, Khaneghah AM, Sant'Ana AS, Eds. Essential Oils in Food Processing Chemistry: Safety and Applications. USA: John Wiley Sons Ltd 2017; pp. 21-55.

[9]   Bruni R, Medici A, Andreotti E, *et al.* Chemical composition and biological activities of Ishpingo essential oil, a traditional Ecuadorian spice from *Ocotea quixos* (Lam.) Kosterm. (Lauraceae) flower calices. Food Chem 2004; 85(3): 415-21.
[http://dx.doi.org/10.1016/j.foodchem.2003.07.019]

[10]   Bastidas V, Mora-Vivas F. Analysis of the essential oil of the leaves of *Guarea guidonia* (L.) Sleumer (Meliaceae). Rev Fac Farm Univ Los Andes 2014; 56(1): 18-20.

[11]   Candan F, Unlu M, Tepe B, *et al.* Antioxidant and antimicrobial activity of the essential oil and methanol extracts of *Achillea millefolium* subsp. millefolium Afan. (Asteraceae). J Ethnopharmacol 2003; 87(2-3): 215-20.
[http://dx.doi.org/10.1016/S0378-8741(03)00149-1] [PMID: 12860311]

[12]   Caroprese Araque JF, Parra Garcés MI, Arrieta Prieto D, Stashenko E. Anatomía microscópica y metabolitos secundarios volátiles en tres estadios del desarrollo de las inflorescencias de *Lantana camara* (Verbenaceae). Rev Biol Trop 2011; 59(1): 473-86.
[PMID: 21516661]

[13]   Grajales-Conesa J, Aceves-Chong L, Rincón-Rabanales M, Cruz-López L. *Jatropha curcas* flowers from southern Mexico: chemical profile and morphometrics. Rev Mex Biodivers 2016; 87(4): 1321-7.
[http://dx.doi.org/10.1016/j.rmb.2016.08.001]

[14]   Paduch R, Kandefer-Szerszeń M, Trytek M, Fiedurek J. Terpenes: Substances useful in human healthcare. Arch Immunol Ther Exp 2007; 55(5): 315-27.
[http://dx.doi.org/10.1007/s00005-007-0039-1] [PMID: 18219762]

[15]   Lopes AP, de Oliveira Castelo Branco RR, de Alcântara Oliveira FA, *et al.* Antimicrobial, modulatory, and antibiofilm activity of *tt*-farnesol on bacterial and fungal strains of importance to human health. Bioorg Med Chem Lett 2021; 47: 128192.
[http://dx.doi.org/10.1016/j.bmcl.2021.128192] [PMID: 34118413]

[16]   Kim C, Hesek D, Lee M, Mobashery S. Potentiation of the activity of β-lactam antibiotics by farnesol and its derivatives. Bioorg Med Chem Lett 2018; 28(4): 642-5.
[http://dx.doi.org/10.1016/j.bmcl.2018.01.028] [PMID: 29402738]

[17]   Berghe D, Vlietinck A, Dey P, *et al.* Vanden Screening methods for antibacterial agents from higher plants. London: Academic Press 1991; pp. 47-69.

[18]   Montero-Recalde M, Vayas L, Avilés-Esquivel D, Pazmiño P, Erazo-Gutierrez V. Evaluación de dos métodos para medir la sensibilidad de inhibición de crecimiento de la cepa certificada de *Staphylococcus aureus* subsp. aureus. Rev Investig Vet Peru 2018; 29(4): 1543-7.
[http://dx.doi.org/10.15381/rivep.v29i4.15185]

[19]   Lewis JS, Ed. Performance Standards for Antimicrobial Susceptibility Testing; Twentieth

Informational Supplement. Wayne, USA: Clinical and Laboratory Standards Institute (CLSI) 2012; pp. 294-369.

[20]   Koneman G, Procop D, Church G, *et al.* Koneman Diagnóstico Microbiológico: Texto y Atlas / Koneman Diagnóstico Microbiológico: Texto y Atlas. Barcelona: Wolters Kluwer 2018.

[21]   Gómez-Sequeda N, Cáceres M, Stashenko EE, Hidalgo W, Ortiz C. Antimicrobial and Antibiofilm Activities of Essential Oils against Escherichia coli O157:H7 and Methicillin-Resistant Staphylococcus aureus (MRSA). Antibiotics (Basel) 2020; 9(11): 730.
[http://dx.doi.org/10.3390/antibiotics9110730] [PMID: 33114324]

[22]   Available from: http://coesant-seimc.org/documents/M%C3%A9todosEspeciales_Sensibilidad.pdf

[23]   Avila JG, de Liverant JG, Martínez A, *et al.* Mode of action of *Buddleja cordata* verbascoside against *Staphylococcus aureus.* J Ethnopharmacol 1999; 66(1): 75-8.
[http://dx.doi.org/10.1016/S0378-8741(98)00203-7] [PMID: 10432210]

[24]   Vargas V, Cisneros C, Durán DA. Bioestadística México: . México: Facultad de Estudios Superiores Iztacala. Universidad Nacional Autónoma de México 2004; p. 220.

[25]   Morris GM, Huey R, Lindstrom W, *et al.* AutoDock4 and AutoDockTools4: Automated docking with selective receptor flexibility. J Comput Chem 2009; 30(16): 2785-91.
[http://dx.doi.org/10.1002/jcc.21256] [PMID: 19399780]

[26]   Adasme MF, Linnemann KL, Bolz SN, *et al.* PLIP 2021: Expanding the scope of the protein–ligand interaction profiler to DNA and RNA. Nucleic Acids Res 2021; 49(W1): W530-4.
[http://dx.doi.org/10.1093/nar/gkab294] [PMID: 33950214]

[27]   Šobotník J, Hanus R, Kalinová B, *et al. (E,E)*-alpha-farnesene, an alarm pheromone of the termite *Prorhinotermes canalifrons.* J Chem Ecol 2008; 34(4): 478-86.
[http://dx.doi.org/10.1007/s10886-008-9450-2] [PMID: 18386097]

[28]   Huelin FE, Murray KE. Alpha-farnesene in the natural coating of apples. Nature 1966; 210(5042): 1260-1.
[http://dx.doi.org/10.1038/2101260a0] [PMID: 5967802]

[29]   Semighini CP, Hornby JM, Dumitru R, *et al.* Farnesol-induced apoptosis in *Aspergillus nidulans* reveals a possible mechanism for antagonistic interactions between fungi. Mol Microbiol 2006; 59(3): 53-64.

[30]   Semighini CP, Murray N, Harris SD. Inhibition of *Fusarium graminearum* growth and development by farnesol. FEMS Microbiol Lett 2008; 279(2): 259-64.
[http://dx.doi.org/10.1111/j.1574-6968.2007.01042.x] [PMID: 18201191]

[31]   Jabra-Rizk MA, Shirtliff M, James C, Meiller T. Effect of farnesol on *Candida dubliniensis* biofilm formation and fluconazole resistance. FEMS Yeast Res 2006; 6(7): 1063-73.
[http://dx.doi.org/10.1111/j.1567-1364.2006.00121.x] [PMID: 17042756]

[32]   Koo H, Rosalen PL, Cury JA, Park YK, Bowen WH. Effects of compounds found in propolis on *Streptococcus mutans* growth and on glucosyltransferase activity. Antimicrob Agents Chemother 2002; 46(5): 1302-9.
[http://dx.doi.org/10.1128/AAC.46.5.1302-1309.2002] [PMID: 11959560]

[33]   Brehm-Stecher BF, Johnson EA. Sensitization of *Staphylococcus aureus* and *Escherichia coli* to antibiotics by the sesquiterpenoids nerolidol, farnesol, bisabolol, and apritone. Antimicrob Agents Chemother 2003; 47(10): 3357-60.
[http://dx.doi.org/10.1128/AAC.47.10.3357-3360.2003] [PMID: 14506058]

[34]   Garvey MI, Piddock LJV. The efflux pump inhibitor reserpine selects multidrug-resistant *Streptococcus pneumoniae* strains that overexpress the ABC transporters PatA and PatB. Antimicrob Agents Chemother 2008; 52(5): 1677-85.
[http://dx.doi.org/10.1128/AAC.01644-07] [PMID: 18362193]

[35]   Wegener A, Damborg P, Guardabassi L, *et al.* Specific staphylococcal cassette chromosome *mec* (SCC *mec*) types and clonal complexes are associated with low-level amoxicillin/clavulanic acid and cefalotin resistance in methicillin-resistant *Staphylococcus pseudintermedius*. J Antimicrob Chemother 2020; 75(3): 508-11.
[http://dx.doi.org/10.1093/jac/dkz509] [PMID: 31846043]

[36]   Akiyama H, Oono T, Huh WK, *et al.* Actions of farnesol and xylitol against *Staphylococcus aureus*. Chemotherapy 2002; 48(3): 122-8.
[http://dx.doi.org/10.1159/000064916] [PMID: 12138327]

[37]   Inoue Y, Shiraishi A, Hada T, Hirose K, Hamashima H, Shimada J. The antibacterial effects of terpene alcohols on *Staphylococcus aureus* and their mode of action. FEMS Microbiol Lett 2004; 237(2): 325-31.
[http://dx.doi.org/10.1111/j.1574-6968.2004.tb09714.x] [PMID: 15321680]

[38]   Koo H, Hayacibara MF, Schobel BD, *et al.* Inhibition of *Streptococcus mutans* biofilm accumulation and polysaccharide production by apigenin and *tt*-farnesol. J Antimicrob Chemother 2003; 52(5): 782-9.
[http://dx.doi.org/10.1093/jac/dkg449] [PMID: 14563892]

CHAPTER 4

Application of Viruses as Carriers in Biotechnology

Viridiana R. Escartín-Alpizar[1,*], Julieta Orozco-Martínez[1] and **Israel Valencia Quiroz[2]**

[1] *Laboratory of Natural Products Bioactivity, UBIPRO, Superior Studies Faculty (FES)-Iztacala, National Autonomous University of Mexico (UNAM), Tlalnepantla de Baz, México State, 54090, México*

[2] *Phytochemistry Laboratory, UBIPRO, Superior Studies Faculty (FES)-Iztacala, National Autonomous University of Mexico (UNAM), Tlalnepantla de Baz, México State, 54090, México*

Abstract: Currently, the development of new vaccine technologies for the treatment of diseases is vital. The use of biotechnology in the application of viruses for the development of vaccines is a relatively new research platform. Viruses have become an important tool in biotechnology, and they are being used in the development of vaccines and anticancer drugs. Some of the viral vectors commonly used to develop vaccines are adenoviruses, adeno-associated viruses, herpes simplex viruses, retroviruses and lentiviruses, among others. Viral vectors have been used as vaccines against a variety of infectious diseases, such as COVID-19, influenza, HIV and malaria. Viruses have also been used to target drugs to cancer cells by using engineered viral vectors that can selectively target and infect cancer cells. In this way, viral vectors can also be used to deliver antitumor drugs. This will selectively target cancer cells. Thus, vectors can be used to deliver therapeutic drugs directly to the tumor, resulting in reduced side effects and improved efficacy.

Keywords: Anticancer viral vectors, Biotechnology, Carriers, Virus-based delivery systems, Viral vectors, Vaccines.

INTRODUCTION

Small infectious agents called viruses penetrate and spread within living cells. They are made of genetic material and protein coats [1], as shown in Fig. (**1**). The definition of a virus is an infectious entity that lacks several essential life activities, such as metabolism or reproduction [2, 3].

---

* **Corresponding author Viridiana R. Escartín-Alpizar:** Laboratory of Natural Products Bioactivity, UBIPRO, Superior Studies Faculty (FES)-Iztacala, National Autonomous University of Mexico (UNAM), Tlalnepantla de Baz, México State, 54090, México; Tel: +525572731888; E-mail: viridianaescartin11d@gmail.com

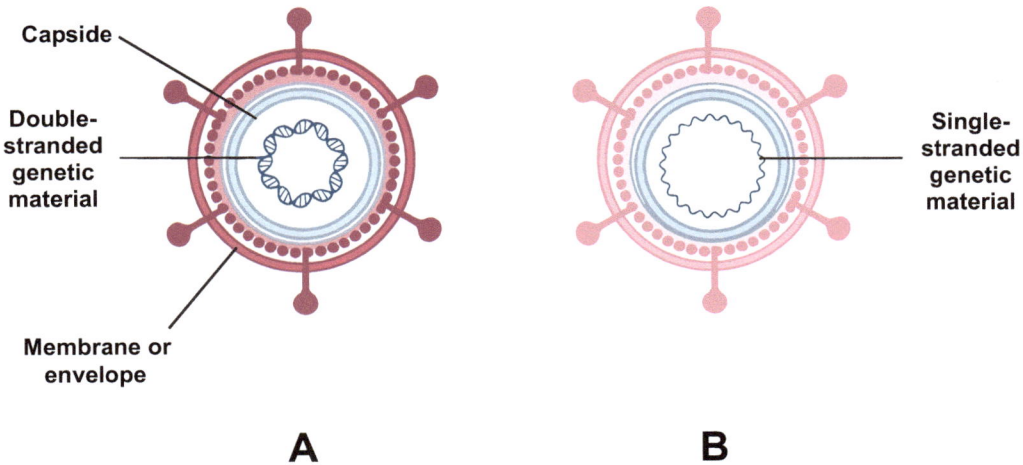

Fig. (1). Basics of virus structure. (**A**) Double-stranded genetic material, and (**B**) Single-stranded genetic material.

To complete their life cycles, viruses might use or abrogate specific cellular machinery components [4, 5]. A replicon, which permits genome duplication, and a capsid, a complex structure that not only protects the viral genome in the extra-cellular space but also plays a role in how virions enter and exit cells, are the two "organs" that the simplest viruses express [3].

In fact, viruses protect their genetic material by enclosing the viral nucleic acid in a protein shell known as a capsid [6], and their capsid proteins associate to produce globally stable structures through locally weak interactions [6]. Genetic engineering can be employed to modify the genetic makeup of the protein capsid [7].

Host ribosomes are enlisted by viruses to translate viral mRNAs. The objective is to ensure that cellular ribosomes are recruited to viral mRNAs, regardless of whether their genomes are made of RNA or DNA or how their mRNA is produced [8]. Host cells may eventually burst as they gather newly generated virus particles. This causes progeny virions to be released, which can subsequently infect other cells and continue the infectious cycle [9]. In tolerant cells, approximately 10,000 progeny virions are created [10]; however, many deficiencies can affect steps between transcription and the formation of new virions as well. Target cells could be infected in the first cycle, but progeny virions would fail to form or be released. Additionally, defective genomes could be incorporated into particles with functional proteins [9].

The host cell may eventually experience lysis when offspring virions accumulate during viral replication. This will release the newly created viral particles, which can then infect nearby cells and continue the infectious cycle [11]. Virus-induced cell lysis releases up to 10,000 virions from a single cell [12]. One mechanism for controlling the generation of mutations in viral populations is thought to be variations in lysis time [5].

Because of their distinct qualities, viruses cannot be categorized as either living things or inert objects [13]. Although they have genetic material and are capable of reproducing, they require a host cell to carry out their basic life functions [3].

Viral nanomaterials can be utilized to transport drugs and genes that can be used to treat and prevent a variety of diseases [7]. The utilization of engineered virus-based nanomaterials as three-dimensional scaffold materials for diagnostic and therapeutic delivery systems as well as technological equipment has greatly advanced recently, demonstrating that viruses can also have advantageous properties [7].

New vaccines based on live recombinant vectors may induce robust, long-lasting immune responses while maintaining acceptable safety profiles [14]. Inducing strong immune responses against the encoded target antigen is possible with viral vector-based vaccinations. Indeed, numerous clinical studies have shown how effective viral vector-based vaccines such as VSV-ZEBOV are at eliciting protective responses in people. However, antigen distribution in a virus-unrelated situation makes this technique extremely difficult to execute [15].

Oncolytic virotherapy is a technique for directly delivering anticancer medications to malignant cells through the use of engineered viruses that can stop the growth of these cells or even kill them [16, 17]. The delivery of the virus to the tumor and the propagation of the virus infection are the two critical phases of effective oncolytic virus therapy, both of which are active areas of preclinical research innovation [18].

In addition, viruses are used to deliver therapeutic drugs since they can infect tumor cells [19]. Adenoviruses, adeno-associated viruses, herpes simplex viruses, retroviruses, lentiviruses, paramyxoviruses, and poxviruses are a few examples of viral vectors frequently employed in biotechnology. Adenoviruses are DNA viruses that do not integrate into the host genome, whereas retroviruses are RNA viruses that integrate their genome into the DNA of the host cell. Herpesviruses are large DNA viruses that cause lifelong latent infections in their hosts, whereas adeno-associated viruses are small DNA viruses that require a helper virus for reproduction [20]. Although not exclusively, many naturally occurring viruses show a preference for tumors and tumor cells. Since most cancers have evolved to evade immune identification and killing as well as to withstand apoptosis and translational suppression, which are the main responses utilized by normal cells to restrict a virus infection, this most likely has more to do with tumor biology than virus biology. Several cytotoxic immune effector mechanisms, including direct virus-mediated cytotoxicity, can be used by oncolytic viruses to kill cancer cells after infection [18].

Indeed, because of their ability to target and deliver genes or medications to particular cells and tissues, viruses have become an important tool in biotechnology. These characteristics allow for their application in the development of vaccinations to prevent various diseases as well as the formulation of effective anticancer treatments [21, 16]. Oncolytic therapies take advantage of viral replication and the activation of the immune system against infected cells, two factors that gene therapy approaches often try to counteract [22]. We may anticipate the production of new and improved approaches using virus-based delivery systems to cure an expanding spectrum of diseases and provide improved protection against them as this technology continues to advance [23, 24]. To construct appropriate clinical trial designs, dosing regimens, pharmacodynamic assays, educational programs addressing biosafety concerns, and novel manufacturing and regulatory channels are necessary for the development of oncolytic viruses as therapeutic medicines [17].

GENERAL ADVANTAGES OF VIRUS-BASED DELIVERY SYSTEMS

Viruses provide a number of benefits over traditional medicine delivery methods, including the ability to target particular cells and tissues [18] and longer-lasting treatment outcomes [19, 25].

In fact, viruses can control the release of both medications and genes, which is a fundamental property that makes them useful for gene therapy and drug delivery [19]. For instance, in the field of gene delivery, viruses have been used to transmit nucleases, short interfering RNAs, and full-length DNA as well as machinery for genome editing [7]. Viruses are used in medication delivery to transport therapeutic cargo for photodynamic treatment (PDT), photothermal therapy (PTT), and chemotherapy [7].

Vaccinia virus and other poxviruses produce a number of proteins that aid in the ability of the virus to thwart the host defenses against infection. Examples include proteins that resist apoptosis, capture chemokines, synthesize steroids, counteract complement activation, interfere with interferon and intercept interleukins [26].

ADVANTAGES OF USING VIRAL VECTORS IN VACCINE DEVELOPMENT

In recent decades, vaccinations against infectious diseases, such as COVID-19, influenza, and HIV, have been created using viral vectors [27]. Adenovirus vector vaccines, for example, use human cells to produce the target antigen. However, they carry a DNA payload that is first translated into an mRNA before becoming the spike protein, as in the case of the COVID-19 vaccine [28]. Due to their immunogenicity and protective effectiveness in rigorous nonhuman primate challenge trials, novel serotype adenovirus (Ad) and cytomegalovirus (CMV) vaccination vectors have emerged over the past several years as promising vectors for HIV-1 and other pathogens [29]. This is because viruses can efficiently induce cellular and humoral immune responses [30]. Compared to conventional vaccine development techniques, recombinant viral vectors have been shown to generate a more potent and discernible immune response [20]. An adjuvant-like effect can be produced by the inherent capacity of adenoviral vectors to stimulate the innate immune response in such a complex manner to improve the process of antigen presentation. This encourages the development of potent humoral and cell-mediated immune responses against the vaccine antigen that would otherwise be less immunogenic [31].

It is possible to modify current delivery systems to better target particular cell types [32]. This technology can be used to create potent anticancer medications as well as vaccinations that can specifically target specific diseases. The deletion of viral genes necessary for virus replication in normal cells but unnecessary in tumor cells, for instance, is one method for achieving the cancer-targeting mechanism of oncolytic viruses.

Another method for achieving tumor-specific targeting of oncolytic viruses is the use of tissue/tumor-specific promoters for critical viral genes. To accomplish this, the virus must be altered to include promoters that are only active in tumor cells. This will cause the viral genes to be activated and cause the virus to replicate exclusively in tumor cells. To further improve tumor-specific targeting of oncolytic viruses, several molecular steps/regulators of the cell cycle can be targeted. This entails altering the virus to specifically target molecules that are overexpressed or dysregulated in tumor cells, which will ultimately result in the virus only replicating in the tumor cells [33].

Indeed, viral vectors can be modified to transport a range of compounds with various properties, increasing their adaptability and effectiveness as a tool in the production of vaccines [24, 23]. A viral vector is used for viral delivery (for example, adeno-associated viruses (AAVs), lentiviruses and adenoviruses) to encapsulate a therapeutic gene in RNA or DNA form to enable effective delivery [34].

The use of nonhuman lentiviruses and herpesviruses as vectors, altering the viral genome to minimize immunogenicity, and creating tactics to target particular cell types have all been suggested as ways to make viral vectors safer. These techniques seek to improve the safety profile of viral vectors for use in gene therapy [12]. They also have many other benefits. First, compared to other approaches, viral vectors have the capacity to transfer substantial amounts of genetic material, permitting a more focused and efficient immune response [35]. Second, viral vectors frequently have their own promoter regions, facilitating the quick and effective production of the delivered gene payload [19, 36]. Some viral vectors can trigger a more thorough immune response, improving infection resistance [37]. Finally, viral vectors can be modified to specifically target certain cells, increasing their effectiveness in eliciting an immune response [23].

When compared to conventional vaccine techniques, virus-based vaccine delivery systems have advantages including cost-effectiveness and efficiency, which can result in a faster and more affordable production process [38]. Technologies from rapid-response manufacturing platforms are used to produce vaccinations at an unparalleled speed and scale for pandemic response [28]. With advantages including accuracy, efficacy, and the ability to transfer substantial amounts of genetic information, viral vectors have shown considerable potential for the development of vaccines and therapeutic interventions [19, 25]. These vectors are considered a potential and cutting-edge tool in the fight against infectious illnesses because they offer an alluring substitute to conventional platforms for delivering vaccine antigens and for precisely targeting and destroying tumor cells [39].

Viruses to Develop Vaccines for COVID-19

By creating prophylactic vaccines and repurposing current medications as potential cures, the scientific and medical communities are making major attempts to restrict the pandemic and subsequent waves of viral propagation [40]. To introduce genetic material into the body and trigger an immunological response, a customized virus can be used [41]. This approach has been demonstrated to be successful since it can elicit both humoral and cellular immunity, assisting in protecting people against the virus [42].

Certain viruses have been shown to function as vectors for the delivery of genetic material, which can result in the induction of immunity against subsequent infections [43]. This method has grown more important as new vaccines and immunotherapies are developed.

A *via*ble alternative to traditional vaccine techniques is the DNA vaccination strategy, which uses genetic material precisely tailored to trigger an immune response against viral proteins [44]. The DNA vaccination method is a *via*ble option for developing future vaccines against a wide range of viruses because it has been successfully used in the development of animal and clinical trials [45].

In general, viruses are proving to be incredibly useful instruments for creating vaccinations to stop the COVID-19 pandemic. These viruses assist the development of immunity against the virus and prevent further infections by using their unique delivery capabilities [46].

Viruses to Develop Influenza Vaccines

Large amounts of genetic information can be delivered by viruses to protect against influenza viruses [47]. AAV-based vectors expressing the A/Mexico/4603/2009 (H1N1) hemagglutinin, nucleocapsid, and the matrix protein M1 induced strong cellular and humoral immune responses in mice [48]. Even in extremely young and old mice, mRNA vaccinations produce balanced, durable, and protective immunity against influenza A virus infections [49]. Furthermore, viruses can be used to distribute single-stranded RNA (ssRNA) molecules, which can stimulate the synthesis of influenza-specific proteins [50]. This method has been used to create influenza vaccines because ssRNA can trigger an immune response against the virus and thus offer protection against illness [51]. The potential for using a recombinant parainfluenza virus 5 incorporating NP from H5N1 as a universal influenza virus vaccine is currently being investigated [52].

In general, viruses can be used to create vaccines to fight influenza. It has been shown that using viruses to transfer genetic material is efficient because they can trigger an immune response that protects against the virus and lasts for a long time [53].

Viruses to Develop Vaccines for HIV

To create vaccinations for HIV, the cause of AIDS, viruses have been studied [29]. The ability of viruses to transmit genetic information to target cells makes them a promising method for creating vaccines that can trigger an immune reaction against the virus [54]. HIV-specific genes have been delivered *via* adeno-associated viruses. This method is considered a potential strategy for combating the virus and has been successfully used in the creation of HIV vaccines [55]. In addition, viruses can be used to deliver HIV-specific small interfering RNAs (siRNAs) as a possible method for locating and silencing the viral genome [56]. Individuals can then be protected from infection using mRNA vaccines, a type of vaccination that uses a small fragment of genetic material from the virus to instruct cells in the body to manufacture a protein that prompts an immune response. Both animal models and humans have responded favorably to these mRNA vaccines, and subsequent technological developments have essentially resolved the problems of mRNA instability and ineffective *in vivo* administration [43]. Additionally, some viruses, including poxviruses, can trigger a thorough immune response that provides improved protection against HIV [57].

In general, viruses can be used in the creation of HIV vaccines. HIV-specific genetic material and small interfering RNAs (siRNAs), which can aid in triggering an immune response against the virus, can be delivered through the use of viral vectors. Therefore, viruses are a crucial tool in the ongoing war against HIV [58].

Viruses to Develop Vaccines for Malaria

Malaria is a potentially fatal disease brought on by Plasmodium and spread to people by specific mosquito species [59]. There were 193.9 million cases of malaria worldwide in 2017 [60]. It mostly occurs in tropical nations. Malaria transmission also occurs in major parts of South Asia, Latin America, and Oceania. *Plasmodium falciparum* infection is the cause of the highest prevalence of the disease in Africa [59]. *P. vivax*, which accounts for up to 65% of infections in the Americas, is also common in these locations in addition to *P. falciparum* [61].

The complicated life cycle and wide range of antigenic variation of the parasite are significant barriers to creating an effective malaria vaccine [62]. Recombinant proteins have traditionally been used as the foundation for the development of malaria vaccines [63].

Viral-vectored vaccines are a highly promising tool for removing the barriers to the eradication of malaria [64]. Using viral vectors to trigger cellular immunity and humoral responses can help treat a wide range of infectious disorders, including certain cancers and malaria [65]. In both animal models and human studies, some possible vaccine candidates have shown the ability to induce malaria-specific immunity [66].

The best way to stimulate cellular immunity is still with viral vector vaccines, which are also now showing promise for stimulating potent humoral responses. This indicates that viral vectors can induce T-cell-mediated immunity as well as antibody production without the use of an adjuvant [65]. Since the antigenic targets for these vaccines are produced by the eukaryotic cellular machinery, it might be conceivable to produce antigens with a native conformation [64].

Finally, some viral vectors have the ability to transmit several genes. Consequently, a single virally vectored construct may have numerous antigens from various parasite life stages and might potentially trigger strong protective immunity [67].

Ability of Virus to Stimulate a Humoral Response

Viruses can be utilized to trigger a target-specific adaptive humoral immune response [68]. The coding of HIV proteins and other antigens is possible using this method. It is possible to alter viral vectors to improve their safety and enhance their immunogenicity [69]. The immune system will then detect the viral vector, leading to the production of an antibody response. When the antigens of the desired gene are recognized and bound by the antibodies generated, infection resistance is achieved [70].

The best method for producing cellular immunity is still using viral vector vaccines, which are also now showing promise for producing potent humoral responses [68]. Poxviruses are a particular class of viruses that have been utilized in studies to stimulate a wide range of immunological responses in people. It has been discovered that this immune response offers higher protection against HIV. Compared to conventional vaccine techniques, this strategy has the potential to generate a more robust and durable immune response against HIV [71].

Ability of Virus to Stimulate a Cellular Immune Response

One of the best methods for producing cellular immunity is still using viral vector vaccines, which are also now showing promise for producing potent humoral responses [68]. This is accomplished by employing viral vectors, such as those that code for HIV proteins, to convey the appropriate gene or antigen [69]. The viral vector can be used to express the desired gene once it enters the cells. This will be detected by the immune system and cause the proper cellular reaction [72]. While cytokines are proteins that help regulate the immune response, therapeutic vaccines increase the immune response against HIV. These interventions, along with other immunological agents, such as human broadly neutralizing antibodies, could be used in combination to effectively suppress the virus [73].

Anticancer Drug Delivery using Viral Vectors

Engineered viral vectors that can specifically target and infect cancer cells are being used to deliver medications to cancer cells utilizing viruses. To infect tumor cells that overexpress the desired receptor, for instance, researchers genetically engineered the measles virus to show a variety of polypeptide ligands on its surface [18]. Despite the effectiveness of local administration, widespread metastatic illness needs systemic therapy. However, natural blood barriers, such as complement and antibodies, limit how often oncolytic viruses can be administered intravenously [74].

In vivo gene therapy employs adeno-associated viruses (AAVs) as vectors for gene transfer. However, their broad tropism restricts their capacity for targeted gene delivery [75]. A number of cytotoxic proteins that are selectively produced in cancer cells have also been added to poxviruses, improving their ability to treat disease. Additionally, efforts are being made to limit or change the tendency of viruses to infect particular cell types [76].

Development of Personalized Vaccines for Cancer Immunotherapy

Research is now being conducted on the application of personalized vaccinations for cancer immunotherapy. Vaccines can be prophylactic (preventative) or therapeutic. Prophylactic vaccines have been shown to be effective in avoiding viral malignancies where the etiology is known, such as hepatitis B virus and human papillomavirus (HPV). However, it is questionable whether endogenous or exogenous vaccines for cancer treatment would be preferred [77]. Personalized vaccines are intended to stimulate the patient's own immune system to identify and target tumor cells in a specific way [78].

Different methods can be used to create customized vaccines. For instance, dendritic cells are frequently used as adjuvants in cancer immunotherapy to deliver antigens because of their distinct properties [79], as shown in Fig. (**2**). Another potential approach for developing vaccines is the use of autologous tumor cells, which are cancer cells originating from the same person in which they are detected [80].

In this situation, peptides—short chains of amino acids—also have a function. These peptides are carefully chosen in tailored vaccines based on the precise mutations found in a patient's tumor to elicit a specific immune response [81].

Viral vectors also offer a different approach for developing customized vaccines. They function by integrating particular antigens from a patient's tumor or infection into the vector [82].

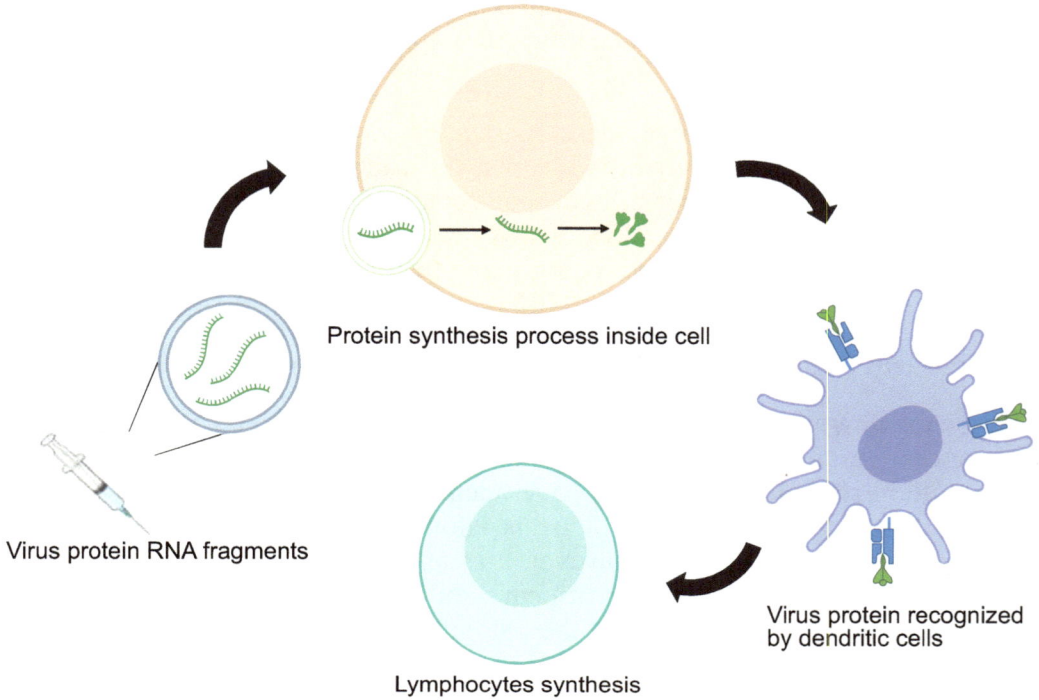

Fig. (2). Diagram illustrating the viral infection and immune response process. First, the image depicts fragments of virus protein RNA, which is the genetic material of the virus. Second, the figure shows the process of protein synthesis inside a cell, where viral protein RNA is used as a template for producing viral proteins. Third, this panel illustrates how the synthesized viral proteins are recognized by dendritic cells, which are crucial components of the immune system. Finally, the final image portrays the stimulation of lymphocyte synthesis, representing the body's immune response to the viral infection.

The effectiveness of personalized vaccines is also being improved by new techniques, such as gene editing. Using CRISPR-Cas9, it was possible to edit the genes of cancer cells and produce a vaccine that specifically targets the patient's tumor-specific mutations. The personalized vaccine was then tested in mice, where it showed encouraging signs of reducing tumor growth [83]. Biomarker-directed immunotherapy uses specialized chemicals called biomarkers, which can be discovered in blood or tumors, to determine the choice of immunotherapy medications. By identifying patients who are most likely to benefit from immunotherapy and choosing the best medication for them, this strategy attempts to customize cancer treatments [84] and strategies for enhancing antigen presentation, such as the use of nanoparticles coated with certain proteins. Stronger immunological responses result from 'the ability of this protein to attach to the antigen and more effectively convey it to the immune system [85].

STRATEGIES USED TO TARGET SPECIFIC CANCER CELLS

Viral vectors and oncolytic viruses are only two of the diverse methods employed to target particular cancer cells. Utilizing viral vectors to deliver therapeutic proteins or medications directly to the tumor can increase their effectiveness and minimize their negative effects [75]. Additionally, oncolytic viruses can be employed to deliver anticancer medications or activate the immune system against cancer cells [18]. Small interfering RNAs (siRNAs) have also been utilized to specifically target tumor cells and this increases efficacy. Nevertheless, the development of clinically appropriate, safe, and efficient drug delivery vehicles is necessary for the widespread use of RNAi therapies for disease prevention and treatment [86].

ONCOLYTIC VIRUSES

Researchers have genetically modified oncolytic viruses to selectively target the tumor cell surface, de-target sensitive tissues, or create dual-target viruses to improve both vascular targeting and tumor infection. Oncolytic viruses can infect and kill cancer cells while leaving normal cells unharmed [18]. These viruses can trigger an immune response against cancer cells and have been utilized to deliver anticancer medications [87]. Additionally, genetically modified transgene-expressing 'armed' oncolytic viruses can be employed to carry genetic material that can alter the expression of genes important in carcinogenesis, improving therapeutic effects (third generation) [88].

Oncolytic adenoviruses use transcriptional targeting as a technique to focus both viral cell lysis and therapeutic gene expression on cancer cells. Tissue-specific promoters are being used to control viral replication to improve therapeutic effects [89]. These promoters can be used with viral vectors to limit virus replication in the appropriate target tissue, improving the safety and specificity of therapeutic strategies [90].

In addition, tissue-specific promoters have been employed to regulate the expression of therapeutic proteins in cancer cells. For instance, the prostate-specific antigen (PSA) promoter has been utilized to drive the expression of a suicide gene in prostate cancer cells, resulting in the selective eradication of these cells [91].

RISKS AND POTENTIAL COMPLICATIONS OF VIRUS-BASED DELIVERY SYSTEMS

Although virus-based delivery systems have many advantages, they can also carry significant dangers and possible problems. Recombinant viral particles, for instance, would not be entirely safe given that some of them might be able to integrate into the host cell genomes [92]. This might result in the expression of unwanted genes and the possibility of negative side effects. Before moving further with any applications, it is crucial to thoroughly weigh the advantages and potential hazards of gene editing [93]. Integration, however, does not ensure sustained transcription because integrated vector genomes have the potential to gradually suppress transgene expression over time [92]. However, careful immunosurveillance carried out as part of ongoing clinical studies will provide the foundation for understanding the nuances of the immune response in AAV-mediated gene transfer, facilitating safe and effective therapies for genetic diseases. Additionally, the safety and efficacy of these systems are still not well defined, and they may result in the development of harmful immune responses [94].

Retroviruses, among other viruses employed in clinical studies based on gene therapy, have also been connected to a higher risk of cancer [95]. This is partly because of their capacity to integrate into the genome of the host, resulting in mutations and chromosomal rearrangements that may eventually result in cancer. Normal cells may transform into tumor cells as a result of genetic alterations caused by mutations in particular genes, such as oncogenes and tumor suppressor genes, or by chromosomal rearrangements [96]. By utilizing approaches such as tissue-specific targeting, gene silencing, and the use of improved delivery systems, these off-target effects can be reduced [97].

Regulatory Issues during Development and Testing of Viral Vectors for Clinical Use

Strict regulatory requirements must be followed during the creation and testing of viral vectors for clinical usage. Several regulatory organizations, notably the US Food and Drug Administration (FDA) and the European Medicines Agency (EMA), oversee the safety and efficacy of gene therapy products. Clinical studies must also follow stringent regulatory requirements, which include obtaining participants' informed consent, monitoring for negative side effects, and reporting any such incidents to regulatory bodies. The regulatory bodies also demand that patients who have had gene therapy be followed up for a lengthy period of time to identify any long-term negative consequences [92].

Viral vectors need to pass the US FDA standards for safety and effectiveness in human use [1]. Additionally, the efficacy and safety profile of the vector should be assessed to ensure that it is successful in delivering the therapeutic gene and has no negative side effects. The risk assessment method should also consider the immunogenicity, efficacy, and safety profile of the vector [98].

CONCLUSION

The advent of using viral vectors as therapeutic tools signifies a transformative shift in the medical landscape, offering outcomes that often surpass traditional methods. As outlined in our discussion, this innovative technique, while promising, is still in its infancy. Comprehensive research is imperative to discern potential side effects and to critically assess its true merits or potential pitfalls. The efficacy of viral vectors, especially in areas like oncology, is undeniable. However, it is equally vital to acknowledge and address the observed challenges that might pose risks to patients already contending with health issues. As such, the ongoing exploration and enhancement of viral vector applications in medical treatments remain of utmost importance.

ACKNOWLEDGEMENT

We would like to express our gratitude to Roman Adrián González Cruz for his invaluable assistance in drawing Fig. (**2**) for this work.

REFERENCES

[1]   Cosar B, Karagulleoglu ZY, Unal S, *et al.* SARS-CoV-2 Mutations and their Viral Variants. Cytokine and Growth Factor Reviews 2022; 63: 10-22.
[http://dx.doi.org/10.1016/j.cytogfr.2021.06.001]

[2]   La Scola B, Desnues C, Pagnier I, *et al.* The virophage as a unique parasite of the giant mimivirus. Nature 2008; 455(7209): 100-4.
[http://dx.doi.org/10.1038/nature07218] [PMID: 18690211]

[3]   Forterre P. Defining life: The virus viewpoint. rig Life Evol Biosph 2010; 40: 151-60.

[4]   Weitzman MD, Fradet-Turcotte A. Virus DNA replication and the host DNA damage response. Annu Rev Virol 2018; 5(1): 141-64.
[http://dx.doi.org/10.1146/annurev-virology-092917-043534] [PMID: 29996066]

[5]   Sanjuán R, Domingo-Calap P. Mechanisms of viral mutation. Cell Mol Life Sci 2016; 73(23): 4433-48.
[http://dx.doi.org/10.1007/s00018-016-2299-6] [PMID: 27392606]

[6]   Zlotnick A. Are weak protein–protein interactions the general rule in capsid assembly? Virology 2003; 315(2): 269-74.
[http://dx.doi.org/10.1016/S0042-6822(03)00586-5] [PMID: 14585329]

[7]   Wen AM, Steinmetz NF. Design of virus-based nanomaterials for medicine, biotechnology, and energy. Chem Soc Rev 2016; 45(15): 4074-126.
[http://dx.doi.org/10.1039/C5CS00287G] [PMID: 27152673]

[8]   Walsh D, Mohr I. Viral subversion of the host protein synthesis machinery. Nat Rev Microbiol 2011; 9: 860-75.
[http://dx.doi.org/10.1038/nrmicro2655]

[9]   Klasse PJ. Molecular determinants of the ratio of inert to infectious virus particles. Prog Mol Biol Transl Sci 2015; 129: 285-326.
[http://dx.doi.org/10.1016/bs.pmbts.2014.10.012] [PMID: 25595808]

[10]   Wold W, Toth K. Adenovirus vectors for gene therapy, vaccination and cancer gene therapy. Curr

Gene Ther 2014; 13(6): 421-33.
[http://dx.doi.org/10.2174/1566523213666131125095046] [PMID: 24279313]

[11]   Wilen CB, Tilton JC, Doms RW. HIV: cell binding and entry. Cold Spring Harb Perspect Med 2012; 2(8): a006866.
[http://dx.doi.org/10.1101/cshperspect.a006866] [PMID: 22908191]

[12]   Kay MA, Glorioso JC, Naldini L. Viral vectors for gene therapy: the art of turning infectious agents into vehicles of therapeutics. Nat Med 2001; 7(1): 33-40.
[http://dx.doi.org/10.1038/83324] [PMID: 11135613]

[13]   Claverie JM, Ogata H. Ten good reasons not to exclude giruses from the evolutionary picture. Nat Rev Microbiol 2009; 7(8): 615-5.
[http://dx.doi.org/10.1038/nrmicro2108-c3] [PMID: 19561626]

[14]   Liniger M, Zuniga A, Naim HY. Use of viral vectors for the development of vaccines. Expert Opin Biol Ther 2014.

[15]   Rauch S, Jasny E, Schmidt KE, Petsch B. New vaccine technologies to combat outbreak situations. Front Immunol 2018; 9: 1963.
[http://dx.doi.org/10.3389/fimmu.2018.01963] [PMID: 30283434]

[16]   Russell SJ, Peng KW, Bell JC. Oncolytic virotherapy. Nat Biotechnol 2012; 30(7): 658-70.
[http://dx.doi.org/10.1038/nbt.2287] [PMID: 22781695]

[17]   Kaufman HL, Kohlhapp FJ, Zloza A. Oncolytic viruses: a new class of immunotherapy drugs. Nat Rev Drug Discov 2015; 14(9): 642-62.
[http://dx.doi.org/10.1038/nrd4663] [PMID: 26323545]

[18]   Ramaj T, Zou X. On the treatment of melanoma: A mathematical model of oncolytic virotherapy. Mathematical Biosciences 2023; 365: 109073.
[http://dx.doi.org/10.1016/j.mbs.2023.109073]

[19]   Ma C C, Wang Z L, X T., He Z Y, Wei Y Q. The approved gene therapy drugs worldwide: from 1998 to 2019. Biotechnol Adv 2020; 40: 1-40.
[http://dx.doi.org/10.1038/83324] [PMID: 11135613]

[20]   Matthews QL, Gu L, Krendelchtchikov A, *et al.* Viral vectors for vaccine development. Nov Gene Ther Approaches 2013.

[21]   Kawai T, Akira S. The role of pattern-recognition receptors in innate immunity: update on Toll-like receptors. Nat Immunol 2010; 11(5): 373-84.
[http://dx.doi.org/10.1038/ni.1863] [PMID: 20404851]

[22]   Naldini L. Gene therapy returns to centre stage. Nature 2015; 526(7573): 351-60.
[http://dx.doi.org/10.1038/nature15818] [PMID: 26469046]

[23]   Naso MF, Tomkowicz B, Perry WL III, Strohl WR. Adeno-Associated Virus (AAV) as a Vector for Gene Therapy. BioDrugs 2017; 31(4): 317-34.
[http://dx.doi.org/10.1007/s40259-017-0234-5] [PMID: 28669112]

[24]   Hasanzadeh A, Hamblin MR, Kiani J, *et al.* Could artificial intelligence revolutionize the development of nanovectors for gene therapy and mRNA vaccines?. Nano Today 2022; 47(7573): 101665.
[http://dx.doi.org/10.1016/j.nantod.2022.101665]

[25]   Wirth T, Parker N, Ylä-Herttuala S. History of gene therapy. Gene 2013; 525(2): 162-9.
[http://dx.doi.org/10.1016/j.gene.2013.03.137] [PMID: 23618815]

[26]   Smith GL. Vaccinia virus immune evasion. Immunol Lett 1999; 65(1-2): 55-62.
[http://dx.doi.org/10.1016/S0165-2478(98)00125-4] [PMID: 10065628]

[27]   Dhama K, Sharun K, Tiwari R, *et al.* COVID-19, an emerging coronavirus infection: advances and prospects in designing and developing vaccines, immunotherapeutics, and therapeutics. Hum Vaccin Immunother 2020; 16(6): 1232-8.

[http://dx.doi.org/10.1080/21645515.2020.1735227] [PMID: 32186952]

[28]   Kis Z, Kontoravdi C, Dey AK, Shattock R, Shah N. Rapid development and deployment of high volume vaccines for pandemic response. J Adv Manuf Process 2020; 2(3): e10060.
[http://dx.doi.org/10.1002/amp2.10060] [PMID: 33977274]

[29]   Barouch DH, Picker LJ. Novel vaccine vectors for HIV-1. Nat Rev Microbiol 2014; 12(11): 765-71.
[http://dx.doi.org/10.1038/nrmicro3360] [PMID: 25296195]

[30]   Choi Y, Chang J. Viral vectors for vaccine applications. Clin Exp Vaccine Res 2013; 2(2): 97-105.
[http://dx.doi.org/10.7774/cevr.2013.2.2.97] [PMID: 23858400]

[31]   Reed S G, Bertholet S, Coler R N, Fiede M. New horizons in adjuvants for vaccine development. Trends Immunol 2013; 30: 23-32.
[http://dx.doi.org/https://doi.org/10.1016/j.it.2008.09.006]

[32]   Yang L, Bailey L, Baltimore D, Wang P. Targeting lentiviral vectors to specific cell types *in vivo*. Proc Natl Acad Sci 2006; 103(31): 11479-84.
[http://dx.doi.org/10.1073/pnas.0604993103] [PMID: 16864770]

[33]   Singh PK, Doley J, Kumar GR, Sahoo AP, Tiwari AK. Oncolytic viruses & their specific targeting to tumour cells. Indian J Med Res 2012; 136(4): 571-84.
[PMID: 23168697]

[34]   Yin H, Kauffman KJ, Anderson DG. Delivery technologies for genome editing. Nat Rev Drug Discov 2017; 16(6): 387-99.
[http://dx.doi.org/10.1038/nrd.2016.280] [PMID: 28337020]

[35]   Kochanek S. Gene transfer with high capacit gutless adenoviral vectors. Gene Funct Dis 2001; 2(2-3): 122-5.
[http://dx.doi.org/10.1002/1438-826X(200110)2:2/3<122::AID-GNFD122>3.0.CO;2-7]

[36]   Nayak S, Herzog RW. Progress and prospects: immune responses to viral vectors. Gene Ther 2010; 17(3): 295-304.
[http://dx.doi.org/10.1038/gt.2009.148] [PMID: 19907498]

[37]   Humphreys IR, Sebastian S. Novel viral vectors in infectious diseases. Immunology 2018; 153(1): 1-9.
[http://dx.doi.org/10.1111/imm.12829] [PMID: 28869761]

[38]   Amanna IJ, Raué HP, Slifka MK. Development of a new hydrogen peroxide–based vaccine platform. Nat Med 2012; 18(6): 974-9.
[http://dx.doi.org/10.1038/nm.2763] [PMID: 22635006]

[39]   Gao J, Mese K, Bunz O, Ehrhardt A. State of the art human adenovirus vectorology for therapeutic approaches. FEBS Lett 2019; 593(24): 3609-22.
[http://dx.doi.org/10.1002/1873-3468.13691] [PMID: 31758807]

[40]   Funk CD, Laferrière C, Ardakani A. A snapshot of the global race for vaccines targeting SARS-CoV-2 and the COVID-19 Pandemic. Front Pharmacol 2020; 11: 937.
[http://dx.doi.org/10.3389/fphar.2020.00937] [PMID: 32636754]

[41]   Voellmy R, Bloom DC, Vilaboa N. A novel approach for addressing diseases not yielding to effective vaccination? Immunization by replication-competent controlled virus. Expert Rev Vaccines 2015; 14(5): 637-51.
[http://dx.doi.org/10.1586/14760584.2015.1013941] [PMID: 25676927]

[42]   Ewer KJ, Lambe T, Rollier CS, Spencer AJ, Hill AVS, Dorrell L. Viral vectors as vaccine platforms: from immunogenicity to impact. Curr Opin Immunol 2016; 41: 47-54.
[http://dx.doi.org/10.1016/j.coi.2016.05.014] [PMID: 27286566]

[43]   Pardi N, Hogan MJ, Porter FW, *et al*. MRNA vaccines a new era in vaccinology. Nat Rev Drug Discov 2018; 17: 261-79.

[44]   Liu MA. DNA vaccines: an historical perspective and view to the future. Immunol Rev 2011; 239(1):

62-84.
[http://dx.doi.org/10.1111/j.1600-065X.2010.00980.x] [PMID: 21198665]

[45]   Liu MA. DNA vaccines: a review. J Intern Med 2003; 253(4): 402-10.
[http://dx.doi.org/10.1046/j.1365-2796.2003.01140.x] [PMID: 12653868]

[46]   Krammer F. SARS-CoV-2 vaccines in development. Nat 2020; 586: 516-27.

[47]   Lee LYY, Izzard L, Hurt AC. A review of DNA vaccines against influenza. Front Immunol 2018; 9:
1568.
[http://dx.doi.org/10.3389/fimmu.2018.01568] [PMID: 30038621]

[48]   Sipo I, Knauf M, Fechner H, *et al.* Vaccine protection against lethal homologous and heterologous
challenge using recombinant AAV vectors expressing codon-optimized genes from pandemic swine
origin influenza virus (SOIV). Vaccine 2011; 29(8): 1690-9.
[http://dx.doi.org/10.1016/j.vaccine.2010.12.037] [PMID: 21195079]

[49]   Petsch B, Schnee M, Vogel AB, *et al.* Protective efficacy of *in vitro* synthesized, specific mRNA
vaccines against influenza A virus infection. Nat Biotechnol 2012; 30(12): 1210-6.
[http://dx.doi.org/10.1038/nbt.2436] [PMID: 23159882]

[50]   Geall AJ, Mandl CW, Ulmer JB. RNA: The new revolution in nucleic acid vaccines. Semin Immunol
2013; 25(2): 152-9.
[http://dx.doi.org/10.1016/j.smim.2013.05.001] [PMID: 23735226]

[51]   Bahl K, Senn JJ, Yuzhakov O, *et al.* Preclinical and Clinical Demonstration of Immunogenicity by
mRNA Vaccines against H10N8 and H7N9 Influenza Viruses. Mol Ther 2017; 25(6): 1316-27.
[http://dx.doi.org/10.1016/j.ymthe.2017.03.035] [PMID: 28457665]

[52]   Li Z, Gabbard JD, Mooney A, *et al.* Single-dose vaccination of a recombinant parainfluenza virus 5
expressing NP from H5N1 virus provides broad immunity against influenza A viruses. J Virol 2013;
87(10): 5985-93.
[http://dx.doi.org/10.1128/JVI.00120-13] [PMID: 23514880]

[53]   Palese P. Influenza: Old and new threats. Nat Med 2004; 10: S82-7.
[http://dx.doi.org/10.1038/nm1141]

[54]   Vasan S, Hurley A, Schlesinger SJ, *et al. In vivo* electroporation enhances the immunogenicity of an
HIV-1 DNA vaccine candidate in healthy volunteers. PLoS One 2011; 6(5): e19252.
[http://dx.doi.org/10.1371/journal.pone.0019252] [PMID: 21603651]

[55]   Daya S, Berns KI. Gene therapy using adeno-associated virus vectors. Clin Microbiol Rev 2008;
21(4): 583-93.
[http://dx.doi.org/10.1128/CMR.00008-08] [PMID: 18854481]

[56]   Westerhout EM, Ooms M, Vink M, Das AT, Berkhout B. HIV-1 can escape from RNA interference
by evolving an alternative structure in its RNA genome. Nucleic Acids Res 2005; 33(2): 796-804.
[http://dx.doi.org/10.1093/nar/gki220] [PMID: 15687388]

[57]   Picker LJ, Hansen SG, Lifson JD. New paradigms for HIV/AIDS vaccine development. Annu Rev
Med 2012; 63(1): 95-111.
[http://dx.doi.org/10.1146/annurev-med-042010-085643] [PMID: 21942424]

[58]   Johnston RE, Johnson PR, Connell MJ, *et al.* Vaccination of macaques with SIV immunogens
delivered by Venezuelan equine encephalitis virus replicon particle vectors followed by a mucosal
challenge with SIVsmE660. Vaccine 2005; 23(42): 4969-79.
[http://dx.doi.org/10.1016/j.vaccine.2005.05.034] [PMID: 16005121]

[59]   Snow RW, Guerra CA, Noor AM, Myint HY, Hay SI. The global distribution of clinical episodes of
Plasmodium falciparum malaria. Nature 2005; 434(7030): 214-7.
[http://dx.doi.org/10.1038/nature03342] [PMID: 15759000]

[60]   Weiss DJ, Lucas TCD, Nguyen M, *et al.* Mapping the global prevalence, incidence, and mortality of

Plasmodium falciparum, 2000–17: A spatial and temporal modelling study. Lancet 2019; 394(10195): 322-31.
[http://dx.doi.org/10.1016/S0140-6736(19)31097-9] [PMID: 31229234]

[61]   Weiss DJ, Lucas TCD, Nguyen M, *et al.* Mapping the global prevalence, incidence, and mortality of *Plasmodium falciparum*, 2000–17: A spatial and temporal modelling study. Lancet 2019; 394(10195): 322-31.
[http://dx.doi.org/10.1016/S0140-6736(19)31097-9] [PMID: 31229234]

[62]   Richie TL, Saul A. Progress and challenges for malaria vaccines. Nature 2002; 415(6872): 694-701.
[http://dx.doi.org/10.1038/415694a] [PMID: 11832958]

[63]   Draper SJ, Sack BK, King CR, *et al.* Malaria vaccines: Recent advances and new horizons. Cell Host Microbe 2018; 24(1): 43-56.
[http://dx.doi.org/10.1016/j.chom.2018.06.008] [PMID: 30001524]

[64]   Hill AVS. Vaccines against malaria. Philos Trans R Soc Lond B Biol Sci 2011; 366(1579): 2806-14.
[http://dx.doi.org/10.1098/rstb.2011.0091] [PMID: 21893544]

[65]   Draper SJ, Heeney JL. Viruses as vaccine vectors for infectious diseases and cancer. Nat Rev Microbiol 2010; 8(1): 62-73.
[http://dx.doi.org/10.1038/nrmicro2240] [PMID: 19966816]

[66]   Ogwang C, Kimani D, Edwards NJ, *et al.* Prime-boost vaccination with chimpanzee adenovirus and modified vaccinia Ankara encoding TRAP provides partial protection against *Plasmodium falciparum* infection in Kenyan adults. Sci Transl Med 2015; 7(286): 286re5.
[http://dx.doi.org/10.1126/scitranslmed.aaa2373] [PMID: 25947165]

[67]   Ewer KJ, Lambe T, Rollier CS, Spencer AJ, Hill AVS, Dorrell L. Viral vectors as vaccine platforms: From immunogenicity to impact. Curr Opin Immunol 2016; 41: 47-54.
[http://dx.doi.org/10.1016/j.coi.2016.05.014] [PMID: 27286566]

[68]   Draper SJ, Heeney JL. Viruses as vaccine vectors for infectious diseases and cancer. Nat Rev Microbiol 2010; 8(1): 62-73.
[http://dx.doi.org/10.1038/nrmicro2240] [PMID: 19966816]

[69]   Barouch DH, Deeks SG. Immunologic strategies for HIV-1 remission and eradication. Science (80-) 2014; 345: 169-74.

[70]   Plotkin SA. Vaccines: past, present and future. Nat Med 2005; 11(S4) (Suppl.): S5-S11.
[http://dx.doi.org/10.1038/nm1209] [PMID: 15812490]

[71]   Buchbinder SP, Mehrotra DV, Duerr A, *et al.* Efficacy assessment of a cell-mediated immunity HIV-1 vaccine (the Step Study): A double-blind, randomised, placebo-controlled, test-of-concept trial. Lancet 2008; 372(9653): 1881-93.
[http://dx.doi.org/10.1016/S0140-6736(08)61591-3] [PMID: 19012954]

[72]   Plotkin SA. Vaccines: past, present and future. Nat Med 2005; 11(S4) (Suppl.): S5-S11.
[http://dx.doi.org/10.1038/nm1209] [PMID: 15812490]

[73]   Pantaleo G, Levy Y. Therapeutic vaccines and immunological intervention in HIV infection. Curr Opin HIV AIDS 2016; 11(6): 576-84.
[http://dx.doi.org/10.1097/COH.0000000000000324] [PMID: 27607591]

[74]   Evgin L, Acuna SA, Tanese de Souza C, *et al.* Complement inhibition prevents oncolytic vaccinia virus neutralization in immune humans and cynomolgus macaques. Mol Ther 2015; 23(6): 1066-76.
[http://dx.doi.org/10.1038/mt.2015.49] [PMID: 25807289]

[75]   Münch RC, Janicki H, Völker I, *et al.* Displaying high-affinity ligands on adeno-associated viral vectors enables tumor cell-specific and safe gene transfer. Mol Ther 2013; 21(1): 109-18.
[http://dx.doi.org/10.1038/mt.2012.186] [PMID: 22968478]

[76]   Thorne SH, Hermiston T, Kirn D. Oncolytic virotherapy: Approaches to tumor targeting and

enhancing antitumor effects. Semin Oncol 2005; 32(6): 537-48.
[http://dx.doi.org/10.1053/j.seminoncol.2005.09.007] [PMID: 16338419]

[77]   Mellman I, Coukos G, Dranoff G. Cancer immunotherapy comes of age. Nature 2011; 480(7378): 480-9.
[http://dx.doi.org/10.1038/nature10673] [PMID: 22193102]

[78]   Sharma P, Allison JP. Immune checkpoint targeting in cancer therapy: toward combination strategies with curative potential. Cell 2015; 161(2): 205-14.
[http://dx.doi.org/10.1016/j.cell.2015.03.030] [PMID: 25860605]

[79]   Palucka K, Banchereau J. Cancer immunotherapy *via* dendritic cells. Nat Rev Cancer 2012; 12(4): 265-77.
[http://dx.doi.org/10.1038/nrc3258] [PMID: 22437871]

[80]   Schumacher TN, Schreiber RD. Neoantigens in cancer immunotherapy. Science 2015; 348(6230): 69-74.
[http://dx.doi.org/10.1126/science.aaa4971] [PMID: 25838375]

[81]   Sahin U, Türeci Ö. Personalized vaccines for cancer immunotherapy. Science 2018; 359(6382): 1355-60.
[http://dx.doi.org/10.1126/science.aar7112] [PMID: 29567706]

[82]   Draper SJ, Heeney JL. Viruses as vaccine vectors for infectious diseases and cancer. Nat Rev Microbiol 2010; 8(1): 62-73.
[http://dx.doi.org/10.1038/nrmicro2240] [PMID: 19966816]

[83]   Stadtmauer EA, Fraietta JA, Davis MM, *et al.* CRISPR-engineered T cells in patients with refractory cancer. Science 2020; 367(6481): eaba7365.
[http://dx.doi.org/10.1126/science.aba7365] [PMID: 32029687]

[84]   Ribas A, Wolchok JD. Cancer immunotherapy using checkpoint blockade. Science 2018; 359(6382): 1350-5.
[http://dx.doi.org/10.1126/science.aar4060] [PMID: 29567705]

[85]   Yewdell JW, Haeryfar SMM. Understanding presentation of viral antigens to CD8+ T cells *in vivo*: the key to rational vaccine design. Annu Rev Immunol 2005; 23(1): 651-82.
[http://dx.doi.org/10.1146/annurev.immunol.23.021704.115702] [PMID: 15771583]

[86]   Whitehead KA, Langer R, Anderson DG. Knocking down barriers: Advances in siRNA delivery. Nat Rev Drug Discov 2009; 8(2): 129-38.
[http://dx.doi.org/10.1038/nrd2742] [PMID: 19180106]

[87]   Andtbacka RHI, Kaufman HL, Collichio F, *et al.* Talimogene laherparepvec improves durable response rate in patients with advanced melanoma. J Clin Oncol 2015; 33(25): 2780-8.
[http://dx.doi.org/10.1200/JCO.2014.58.3377] [PMID: 26014293]

[88]   Liu TC, Galanis E, Kirn D. Clinical trial results with oncolytic virotherapy: A century of promise, a decade of progress. Nat Clin Pract Oncol 2007; 4(2): 101-17.
[http://dx.doi.org/10.1038/ncponc0736] [PMID: 17259931]

[89]   Nettelbeck DM. Cellular genetic tools to control oncolytic adenoviruses for virotherapy of cancer. J Mol Med 2008; 86(4): 363-77.
[http://dx.doi.org/10.1007/s00109-007-0291-1] [PMID: 18214411]

[90]   Yamamoto M, Curiel DT. Current issues and future directions of oncolytic adenoviruses. Mol Ther 2010; 18(2): 243-50.
[http://dx.doi.org/10.1038/mt.2009.266] [PMID: 19935777]

[91]   Duarte S, Carle G, Faneca H, Lima MCP, Pierrefite-Carle V. Suicide gene therapy in cancer: Where do we stand now? Cancer Lett 2012; 324(2): 160-70.
[http://dx.doi.org/10.1016/j.canlet.2012.05.023] [PMID: 22634584]

[92]   Thomas CE, Ehrhardt A, Kay MA. Progress and problems with the use of viral vectors for gene therapy. Nat Rev Genet 2003; 4(5): 346-58.
[http://dx.doi.org/10.1038/nrg1066] [PMID: 12728277]

[93]   Hacein-Bey-Abina S, Von Kalle C, Schmidt M, *et al.* LMO2-associated clonal T cell proliferation in two patients after gene therapy for SCID-X1. Science 2003; 302(5644): 415-9.
[http://dx.doi.org/10.1126/science.1088547] [PMID: 14564000]

[94]   Mingozzi F, High KA. Immune responses to AAV vectors: Overcoming barriers to successful gene therapy. Blood 2013; 122(1): 23-36.
[http://dx.doi.org/10.1182/blood-2013-01-306647] [PMID: 23596044]

[95]   Hacein-Bey-Abina S, Garrigue A, Wang GP, *et al.* Insertional oncogenesis in 4 patients after retrovirus-mediated gene therapy of SCID-X1. J Clin Invest 2008; 118(9): 3132-42.
[http://dx.doi.org/10.1172/JCI35700] [PMID: 18688285]

[96]   Braoudaki M, Tzortzatou-Stathopoulou F. Tumorigenesis related to retroviral infections. J Infect Dev Ctries 2011; 5(11): 751-8.
[http://dx.doi.org/10.3855/jidc.1773] [PMID: 22112727]

[97]   Hajitou A. Targeted systemic gene therapy and molecular imaging of cancer contribution of the vascular targeted AAVP vector. Adv Genet 2010; 69: 65-82.
[http://dx.doi.org/10.1016/S0065-2660(10)69008-6] [PMID: 20807602]

[98]   Müller OJ, Kaul F, Weitzman MD, *et al.* Random peptide libraries displayed on adeno-associated virus to select for targeted gene therapy vectors. Nat Biotechnol 2003; 21(9): 1040-6.
[http://dx.doi.org/10.1038/nbt856] [PMID: 12897791]

CHAPTER 5

Phenolic Compounds with Photo-Chemoprotective Activity

Erick Nolasco-Ontiveros[1,*], María del Socorro Sánchez-Correa[2], José Guillermo Avila-Acevedo[1], Rocío Serrano-Parrales[3] and **Adriana Montserrat Espinosa-González[1]**

[1] *Phytochemistry Laboratory, UBIPRO, Superior Studies Faculty (FES)-Iztacala, National Autonomous University of Mexico (UNAM), Tlalnepantla de Baz, México State, 54090, México*

[2] *Scientific Research Laboratory I, Superior Studies Faculty (FES)-Iztacala, National Autonomous University of México (UNAM), Tlalnepantla de Baz, México State, 54090, México*

[3] *Laboratory of Bioactivity of Natural Products, UBIPRO, Superior Studies Faculty (FES)-Iztacala, National Autonomous University of Mexico (UNAM), Tlalnepantla de Baz, México State, 54090, México*

Abstract: Skin cancer has one of the highest incidence rates among all types of cancer and is predominantly caused by exposure to ultraviolet radiation from the sun, which reaches the Earth's surface due to the well-known phenomenon of thinning of the ozone layer in the stratosphere. To reduce the risk of developing this malignancy, the use of sunscreens is recommended; however, the synthetic compounds in sunscreens can cause side effects and harm the environment. To avoid damage to human health and the environment, the use of different plant secondary metabolites with photochemoprotective potential has been investigated in recent decades. For this reason, phenolic compounds are useful alternatives since many of them are capable of absorbing ultraviolet radiation (UVR). Moreover, some of these compounds have anti-inflammatory, antioxidant, and even anticancer activities. This chapter explores the progress in the study of different phenolic compounds extracted from plants with potential for use in sunscreen formulations.

Keywords: DNA photodamage, Erythema, Flavonoids, Inflammation, Lignins, Natural products, Oxidative damage, Oxidative stress, Phenylpropanoids, Phenolic acids, Photoaging, Photocarcinogenesis, Photochemoprotection, Photoprotection, Reactive oxygen species, Secondary metabolites, Skin cancer, Sunscreens, Tannins, Ultraviolet radiation.

* Corresponding author Erick Nolasco-Ontiveros: Phytochemistry Laboratory, UBIPRO, Superior Studies Faculty (FES)-Iztacala, National Autonomous University of Mexico (UNAM), Tlalnepantla de Baz, México State, 54090, México; Tel +5256231136; E-mail: ericknolasco@iztacala.unam.mx

Israel Valencia Quiroz (Ed.)

INTRODUCTION

Skin cancer has a rate of high incidence in North America, Europe, and Australia, with at least 1.5 million cases reported worldwide in 2020 [1]. In Latin America, cases have increased due to the higher prevalence of fair-skinned individuals, while Africa and Asia have lower incidence rates due to greater skin pigmentation. The International Agency for Research on Cancer, an institution operated by the World Health Organization, has officially classified ultraviolet radiation (UVR) in group I of carcinogens with the strongest evidence of carcinogenicity in humans, especially skin cancer [2].

UVR is in the nonionizing radiation region of the electromagnetic spectrum, representing approximately 9% of the emitted solar radiation. For study purposes, UVR has been divided into three types: UV-C (200-280 nm) the shortest wavelength of radiation; UV-B (280-320 nm) which includes the mid-wavelength range and covers 5% of the total UVR spectrum; and UV-A (320-400 nm), which is known as longwave UV radiation, which comprises about 95% of the total ultraviolet spectrum [3]. The ozone (O_3) present in the planet's stratosphere, also known as the ozone layer, efficiently filters much of the UV-C radiation. Nevertheless, absorption by the ozone layer is rapidly decreasing, particularly the absorption of radiation with wavelengths greater than 280 nm, with an absorption rate of 0% for wavelengths greater than 330 nm [4]. Under normal conditions, the ozone layer filters approximately 80% of UV-B radiation, however, human activities have caused a reduction in the concentration of stratospheric ozone through the emission of compounds such as chlorofluorocarbons (CFCs). This has led to UV-C radiation and a higher percentage of UV-B radiation reaching the Earth's surface, affecting all living organisms. UVR that reaches the Earth varies due to factors such as geography, altitude, ground reflectance, cloud cover, and seasonal variation in ozone concentration in the stratosphere [5].

The mildest skin damage caused by prolonged exposure to UVR is erythema (sunburn), which is characterized by vasodilation and subsequent infiltration of neutrophil leukocytes that activate melanogenesis, the proliferation of epidermal cells, as well as the release of prostaglandins, nitric oxide (NO) and cytokines [6]. One factor that contributes to the development of skin cancer is the accumulation of reactive oxygen species (ROS), which leads to an imbalance in endogenous antioxidant systems, provoking oxidative damage, inflammation, and remodeling of the extracellular matrix [7]. In addition, exposure to UVR-A leads to an increase in ROS within cells, provoking oxidation chain reactions. The damage caused by UVR-B is mediated by ROS and by the increase in reactive nitrogen species, which triggers an increase in peroxidized lipids and a reduction in catalase and glutathione peroxidase levels [8]. Subsequently, ROS attacks lipids, proteins, and DNA. Oxidative damage includes the formation of the 8-oxo-7 and 8-oxo-dG photoproducts that cause transversion mutations during DNA replication, the formation of cyclobutane pyrimidine dimers (CPDs), and inflammatory mediators derived from peroxidized lipids [9].

Inflammation caused by UV radiation is a defense mechanism of the body; however, when inflammation becomes chronic, it can lead to a higher probability of developing skin cancer [10]. ROS act as inducers of proinflammatory genes. Proinflammatory mediators are released from keratinocytes, fibroblasts, tumor cells, leukocytes, and the endothelial lining of blood vessels. Plasma mediators include those of a protein nature, such as bradykinin, plasmin, and fibrin; lipid mediators, such as prostaglandins, leukotrienes, and platelet-activating factors; and the inflammatory cytokines IL-1, IL-6, as well as TNF-α, histamine, and active phospholipase [11, 12].

Although enzymatic and nonenzymatic antioxidant systems are sufficient to decrease the damage caused by oxidative stress, when found at low levels, even small concentrations of ROS can trigger the inflammatory process through the activation of NF-kB and AP-1 [13]. NF-kB is a family of five structurally related proteins that function as central mediators of inflammatory responses [14, 15].

Additionally, UVR causes skin photoaging, which is characterized by fewer collagen fibers and fibroblasts, the loss of elasticity, and a reduction in wound-healing capacity [16]. UVR directly activates cell surface receptors, initiating signaling cascades that increase the transcription of different matrix metalloproteinase (MMPs) genes and reduce the expression of procollagen I and III genes, which regulate collagen synthesis and its subsequent degradation. During the photoaging process, the density of Langerhans cells and T lymphocytes is reduced considerably [16, 17]. These changes increase susceptibility to photocarcinogenesis and chronic skin infections. Clinically, signs of photoaging include wrinkles, irregular pigmentation, dryness, roughness, and a variety of premalignant lesions such as actinic keratoses [18].

At the cellular level, exposure to UVR-A induces DNA modifications, due to the oxidation of guanine to 8-oxo-dG, and the creation of DNA strand breaks [19]. UVR-B more efficiently produces mutations and DNA lesions, such as the formation of CPDs and pyrimidine-pyrimidone (6-4) photoproducts. CPDs are the predominant photoproduct induced by UVR [20, 21]. However, both types of lesions mentioned above are highly mutagenic, and they accumulate due to the excessive rate at which repair mechanisms occur, causing abnormal cell proliferation and tumor development [22].

Photocarcinogenesis is a process that involves the accumulation of CPDs, the suppression of the immune system, the depletion of antioxidant defenses, the increase in prostaglandin synthesis, and the induction of ornithine decarboxylase and cyclooxygenase, all of which occur after prolonged UVR exposure [23, 24]. Photocarcinogenesis consists of three stages: 1) tumor initiation, where UVR induces DNA damage in normal skin cells; 2) tumor promotion, when the clonal expansion of initiated cells takes place, resulting in premalignant cells due to alterations in transduction pathways (this stage is still considered reversible); and 3) tumor progression when papillomas transform into carcinomas [25].

Once developed, skin cancer can be classified based on the cell type of origin: melanomas, which develop from melanocytes, are the most aggressive, while nonmelanoma (CNM) tumors develop from keratinocytes and are divided into basal cell (BCC) and squamous cell (SCC) carcinomas. Although BCC and SCC have low mortality and metastasis rates, they are the most frequent types of CNM and can cause the destruction of adjacent tissue. The main etiological factor of skin cancer is UVR, and the latency period of skin cancer can last for decades, from the first exposure to the sun's rays to manifestation [28].

PHOTOCHEMOPROTECTION

Since the 1970s, sunscreens have been widely manufactured and used as an effective strategy for preventing skin neoplasms [29]. Nevertheless, recently, a popular approach to reduce the harmful effects of UVR, photochemoprotection, has been implemented, which uses organic and inorganic compounds applied to the skin to form a protective barrier [30]. These compounds act as filters, absorbing, reflecting, or scattering light to release lower energy radiation without decay. These compounds can be obtained from natural or synthetic sources.

Sunscreens are classified according to their sun protection factor (SPF), which determines the proportion of UVR needed and the time after which skin burns occur [32]. In the final sunscreen emulsions, the structure and properties of the filter molecules, in addition to the substances used as a vehicle must be considered. Some applied forms of sunscreen, such as gels, emulsions, and oils, have limitations in terms of their absorption, removal, and low stability; however, new carriers such as liposomes and nanoparticles have been proposed [33].

Although the effectiveness of synthetic sunscreens has been proven, there is evidence of their adverse effects and potential risks. For example, according to the United States Centers for Disease Control (CDC), in 2007, high levels of oxybenzone, an active ingredient in synthetic sunscreens whose toxicity has already been reported, were reported in mothers who gave birth to children with low birthweights. On the other hand, according to Australia's Therapeutic Goods

Administration (TGA), zinc oxide and titanium dioxide induce the formation of free radicals when exposed to sunlight. These compounds, used instead of para-aminobenzoic acid (PABA), have been reported to promote the appearance of dermatitis and skin discoloration [34]. For this reason, the use of extracts from plants may be a good option to prepare photoprotective formulations.

In addition to the detrimental effects that some sunscreens have on human health, it is important to note that the presence of photoprotective formulations can also affect ecosystems, especially aquatic ecosystems. Recent work has focused on determining the amounts of these compounds in aquatic environments; for example, the sunscreen 2-ethylhexyl-4-methoxycinnamate (EHMC) has been found in an average concentration of 242 ng/g dry weight of fish from four Spanish rivers [35]. To reduce the harmful effects of this type of compound, the use of plant secondary metabolites as sunscreens has been explored.

Extracts and compounds from plants have been tested in different biological models as inhibitors of carcinogenesis with promising results [36]. For instance, many isolated plant compounds have been shown to have good effects as photochemoprotectors that regulate molecular processes and function as inhibitors of angiogenesis, metastasis, proliferation, and apoptosis induction [37].

Phenolic compounds are among the secondary metabolites or natural products that have a greater capacity to prevent and reduce the damage caused by UVR. Additionally, these compounds possess high antioxidant capacity and can function as physical barriers. Therefore, this chapter explores the photochemoprotective potential of different phenolic compounds.

PHENOLIC COMPOUNDS

Phenols are compounds with at least one phenol group in their structure, and when found in plants, they comprise the most commonly synthesized group of secondary metabolites with the highest accumulation; the diversity of phenolic compounds exceeds 10,000 structures [38]. Plant phenolic compounds provide mechanical protection, they attract pollinators, produce growth inhibitors to compete with other plants, and absorb UVR [39]. Among the main subgroups of phenolic compounds are flavonoids, phenolic acids, stilbenes, coumarins, lignans, and tannins [40].

Phenolic compounds are secondary metabolites with well-documented antioxidant activity (Fig. **1**). These compounds are easily consumed in diets that include fruits and vegetables. However, in regard to topical treatments, it is necessary to consider factors such as hydrophobicity and the design of adequate formulations to induce skin penetration [41].

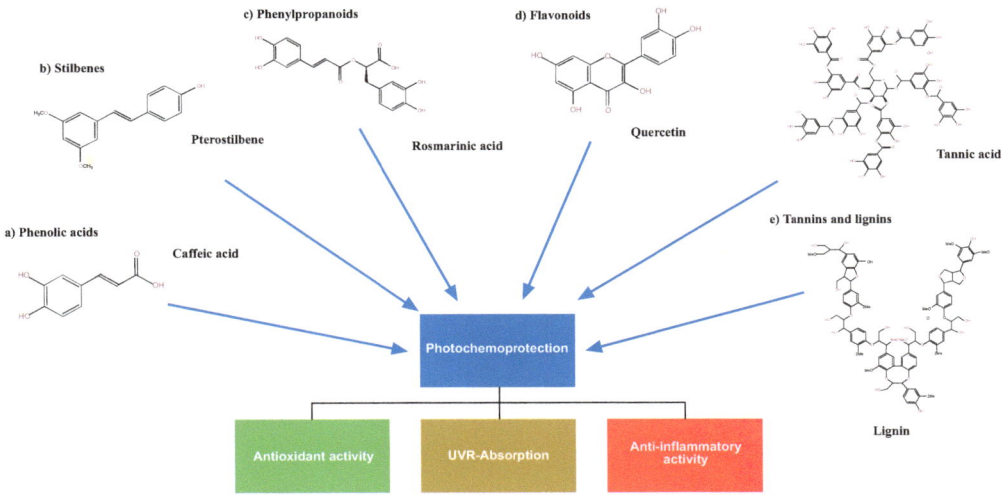

Fig. (1). Families of plant phenolic compounds with photo chemoprotective activity, anti-inflammatory and antioxidant activity, and UVR absorption capacity have been reported.

Phenolics induce cellular defense systems in humans, including antioxidant and detoxifying enzymes, and modulate cell cycle signaling and anti-inflammatory pathways. This leads to cell cycle arrest and apoptosis, with some phenolics promoting apoptosis and others inhibiting angiogenesis [42].

Phenolic Acids

Caffeic Acid

Caffeic acid (8-3,4-dihydrocinnamic acid, (Fig. **1**)) is an acid present in coffee and in a wide variety of fruits and herbaceous plants. Balupillai *et al.* (2018) [43] demonstrated that caffeic acid prevents cyclobutane pyrimidine dimers (CPDs) generation, oxidative DNA damage, ROS formation, and apoptosis in human dermal fibroblasts (HDFs) after controlled exposure to UVR-B. They also found that treatment with caffeic acid inhibits the production of PI3K and AKT, two kinases involved in the development of various types of cancer, including skin cancer, by inducing the expression and activation of the phosphatase and tensin homolog (PTEN), which is a strong suppressor of tumorigenesis. Additionally, caffeic acid promotes the expression of XPC-dependent NER proteins such as XPE, TIFIIH, and ERCC1 *in vitro* and *in vivo*. On the other hand, caffeic acid has also been found to promote a decrease in UVR-A-induced Matrix Metalloproteinase-1 (MMP-1) in HA-Cat keratinocytes by restoring the antioxidant defense system at the cellular and molecular levels [44]. These data demonstrate that caffeic acid prevents UVR-B-induced photo-damage.

Another protective effect of caffeic acid is its ability to reduce UVR-B-induced inflammation and skin carcinogenesis through the activation of peroxisome proliferator-activated receptor γ (PPARγ) in mouse skin [45]. In addition, caffeic acid was found to inhibit STAT-3 translocation, preventing cutaneous photocarcinogenesis in mice [46].

Caffeic acid phenethyl ester (CAPE) is a compound widely distributed in plants that suppresses the growth of melanoma cells. Electroporation of melanoma cells in the presence of CAPE showed high oxidative stress that correlates with high cytotoxicity in this cell type. In addition to inducing apoptosis in cancer cells, this same work suggested that electroporation is a more efficient method for CAPE application, reducing the dose and exposure time needed [47].

Ferulic Acid

Ferulic acid is ubiquitously present in the seeds and leaves of many plant species, in either its free form or covalently conjugated to polysaccharides, glycoproteins, polyamines, lignin, and hydroxy fatty acids of the plant cell wall. This acid absorbs UV radiation and inhibits melanin formation through competitive inhibition with tyrosine, to which it has great structural similarity. The incorporation of ferulic acid in topical formulations with 5% L-ascorbic acid and 1% alpha-tocopherol doubles the photoprotective activity against UVR damage in the long term [48].

Gallic Acid

Since UVR induces the formation of reactive oxygen, the effect of gallic acid on intracellular ROS production was evaluated through flow cytometry. Treatment with 10 micromolar gallic acid after irradiation decreased the formation of ROS by 10 to 45%. Gallic acid also helps to regulate the production of MMP-1 and IL-6 in addition to increasing the secretion of procollagen type 1 in human fibroblasts irradiated with UV-B. Finally, oral administration of gallic acid to hairless mice exposed to UV radiation prevented the formation of wrinkles and dry skin by positively regulating procollagen I and elastin by stimulating TGF-β1 and decreasing MMP-1 and IL-6 levels [49].

Phenylpropanoids

Rosmarinic Acid

Caffeic acid derivative that is produced by several species of the Lamiaceae family [50], the rosmarinic acid molecule efficiently absorbs UVR-B and has been shown to induce anti-inflammatory responses in human keratinocytes exposed to this type of radiation [51]. Pretreating HaCaT cells with rosmarinic acid before exposure to UV-B results in greater cell *via*bility, with reduced DNA fragmentation and apoptotic body formation. In addition, rosmarinic acid contributes to the recovery of Nrf2 levels, as they decrease due to exposure to UV-B (Fernando *et al.*, 2016). Other works have reported that rosmarinic acid has immunomodulatory effects *in vitro* and *in vivo*, reducing the infiltration of mast cells and eosinophils and the concentration of histamine [52].

Stilbenes

Resveratrol

This stilbene has been reported in more than 70 plant species, including grapes, plums, and berries [53]. *In vitro* studies show that the addition of 0.1 μM resveratrol to cell cultures after UV-A irradiation inhibits keratinocyte quenching by almost 100%. Moreover, resveratrol induces an increase in SOD and GSH-Px levels and reduces lipid peroxidation [54].

The antiproliferative and proapoptotic activities of resveratrol include modulating expression of miRNA suppressors, the transcription factors NBkB, PPAR, PGC1α, NRF1, NRF2, and p53, and TGFβ signaling pathways; moreover, its synergistic effects with drugs used in chemotherapy have also been studied [55]. Topical application of this compound to SKH-1 mice significantly inhibited UV-B-induced skin edema, and histological analysis showed that the application of resveratrol causes a significant decrease in hydrogen peroxide generation and leukocyte infiltration [56]. The mechanisms of action of resveratrol against photocarcinogenesis in different biological models are mainly the induction of antioxidant systems, apoptosis, decreased inflammation, cell cycle arrest, and lipid peroxidation [57 - 59].

Piceatannol

Piceatannol is a stable compound that has been reported to have various biological activities, including a strong antioxidant capacity. Maruki-Uchida (2013) [60] demonstrated that piceatannol can decrease the intracellular level of reactive oxygen species in UV-B-irradiated HaCaT-type keratinocytes. The study found that ROS levels were reduced by 13% after treatment with 0.5 μg/ml piceatannol, 21% with 1 μg/ml, and 58% with 2 μg/ml compared to untreated cells. On the other hand, Du *et al.*, (2017) [61] demonstrated that piceatannol inhibits cell growth and induces apoptosis in the melanoma cell lines WM266-4 and A2058. In addition, after measuring the expression levels of 16 miRNAs, it was found that miRNA-181-a expression was significantly higher in cells treated with piceatannol than in controls and even lower in melanoma tissues.

Pterostilbene

Pretreating female SKH1 mice with pterostilbene resulted in a reduction in UVR-B-induced tumorigenesis in 90% of the irradiated mice over 40 weeks. Additionally, it has been reported that stilbene has a higher antimelanogenesis capacity than resveratrol, showing a 63% decrease in melanogenesis with a 58% decrease in intracellular tyrosinase activity in B16F10 cells at a concentration of 10 μM [62].

Flavonoids

Flavonoids are a group of secondary metabolites produced by plants that are characterized by the presence of 15 carbon atoms in their chemical structure, forming 2 benzene rings (A and B) that are attached to a third heterocyclic pyran ring (ring C). The variety of compounds belonging to this family is differentiated by their oxidation levels and by the substitution pattern of the C ring [63]. Because of their chemical structure, flavonoids protect plants from UVR (Fig. **1**); for this reason, flavonoids are good candidates for use as sunscreens.

Flavonoids have shown great potential as protectors against different types of cancer since they present antioxidant, gene regulation, and enzyme inhibition activities and participate in the deregulation of different miRNAs related to the development of cancer [64].

Flavonols

The main sources of flavanols are strawberries, apples, chocolate, cocoa, beans, cherries, and green and black tea. Flavonols have shown potential against the development of oral, rectal, and prostate cancer in humans [65].

Catechin and Epicatechin

Two flavonols present in cocoa, catechin, and epicatechin, have also been shown to have a photoprotective effect against the development of erythema caused by UVR by improving skin conditions. For instance, in a 2006 study [66], women aged 18-65 who consumed either catechin (20 mg/day) and epicatechin (61 mg/day) or equivalent doses of cocoa flavonols for 12 weeks experienced a 15-25% reduction in erythema caused by UVR compared to the control group. Additionally, it was found that the presence of flavonols favored cutaneous blood flow and thus helped maintain skin structure and texture, keeping the skin in good condition.

Epicatechin is also an antioxidant with potential anti-inflammatory effects that is correlated with the suppression of the cyclooxygenase (COX) and lipoxygenase

(LOX) pathways, which are targets of other chemopreventive agents. It has been demonstrated that epicatechin inhibits the growth of human tumor cell lines, including melanoma. On the other hand, in *in vivo* studies in which epicatechin was applied topically to mice, significant reductions were found in both the number and size of UVR-induced skin tumors without toxicity. Finally, this compound significantly inhibits AP-1 transcriptional activation in mouse and human keratinocytes [67].

Flavones

Flavone sources are the Siberian larch tree, onions, milk thistle, acai palm, lemon juice, orange juice, grape juice, cherries, leeks, brussels sprouts, peppers, broccoli, and parsley. Flavones can fight breast, lung, thyroid, stomach, laryngeal, colon, and oral cancer and leukemia [65].

Apigenin

4',5,7-trihydroxyflavone, is a flavone produced abundantly by several plant species (celery, parsley, chamomile, artichokes, and oregano). Apigenin has been shown to possess antioxidant, anti-inflammatory, and anticancer properties. After topical application of apigenin, the appearance of skin cancer induced by UVR can be prevented or the incidence of tumors in SKH-1 mice can be reduced. Apigenin also prevents UV-induced skin tumorigenesis by inhibiting the cyclins and cyclin-dependent kinases that drive cell cycle progression in both mouse keratinocytes and human diploid fibroblasts. Apigenin inhibits UVB-induced cutaneous angiogenesis by maintaining the normally high levels of endogenous thrombospondin 1 (TSP1) by attenuating neoangiogenesis, proliferation, and epidermal thickening in mice exposed to UVR-B [68].

Luteolin

Luteolin has been associated with a wide range of health benefits, including its reported anticancer properties [69]. Luteolin can penetrate the deepest layers of the skin where it has been shown to help regulate aging and inflammation. Additionally, luteolin application helps counteract oxidative stress and provides anti-inflammatory effects in skin carcinoma. Luteolin also has an antiangiogenic effect in retinal neovascularization, due to its antioxidant capacity, which initially decreases ROS production and subsequently affects the proangiogenic function of vascular endothelial growth factor [70].

Quercetin and Rutin

The flavonoids quercetin (3,4-dihydroxyflavonol) and rutin (quercetin-3-O-β-rutinoside) are two of the most studied secondary metabolites. The therapeutic properties of these compounds, mainly quercetin, which is the main flavonoid present in the human diet, have been identified since 1990. These compounds have antioxidant, anticancer, and UVR photoprotective potential [71].

Choquenet *et al.* (2008) [72] evaluated the properties of quercetin and rutin, determining the SPF values of the inorganic sunscreens with which they were combined. Three different combinations were evaluated using the same percentage of each ingredient (10% w/w): flavonoids with TiO_2 and ZnO. Each combination was tested to determine if there were additive or synergistic effects, as well as possible incompatibilities. In the 10% quercetin combined with TiO_2 formulation, SPF values similar to those obtained with homosalate (a compound used as a standard in SPF determination) were obtained. These two flavonoids (quercetin and rutin) provide an adequate level of photoprotection in the UVR-A range.

Quercetin can suppress the activation of STAT3 through IL-6, a protein that promotes cell growth and survival, by reducing the secretion of cyclin D1 and MMP-2. This inhibition leads to a decrease in cell proliferation and migration through modulation of the S and G2/M phases of the cell cycle [73, 74]. Quercetin can also be found in HER-2 phosphorylation, which results in the suppression of the C-Fos promoter and cyclin D1. This enhances cyclin D1 and BCL levels, leading to cell cycle arrest and apoptosis [75]. In a study conducted by Soll *et al.* (2020) [76], quercetin was found to be effective in M16 murine melanoma cells, producing a 75% decrease in cell *via*bility between 6 and 48 hours after treatment. This reduction was comparable to that produced by etoposide, a drug derived from the lignan podophyllotoxin.

The photoprotective effect of rutin has also been tested in combination with anti-inflammatory compounds such as ascorbic acid. Gegotek *et al.* (2020) [77] evaluated the effect of these two compounds on UVR-B irradiated fibroblasts and analyzed the proteomic profile of these cultured cells in a three-dimensional (3D) system. The study revealed inhibition of the proinflammatory signals induced by UVR, and these effects were observed only when the two compounds were used together. The antioxidant effect of rutin and ascorbic acid prevents protein modification and lipid peroxidation. Ascorbic acid stimulates the formation of the protein-rutin adductor, which supports the intra- and extracellular signaling pathway Nrf2/ARE, ultimately regulating antioxidant activity and contributing to protection after UVR exposure.

Kaempferol

Kaempferol (3,5,7-trihydroxy-2-(4-hydroxyphenylchromen-4-one) is one of the flavonoids with the greatest presence in the plant kingdom; it is a tetrahydroxyflavone found in various parts of plants. Kaempferol and its glycosylated derivatives possess cardioprotective, anti-inflammatory, antimicrobial, and antitumor activities [78].

Yao *et al.* (2014) [79], reported that kaempferol suppressed UVR-induced skin carcinogenesis in a mouse model. The data showed that topical application of this compound delayed the growth of tumors with up to 21 weeks of irradiation compared to the control group. Tumor incidence was also reduced by 68 to 91% after 25 weeks of UVR exposure. The inhibitory effects of kaempferol were a direct effect of the inhibition of the MAP kinases: RSK-A2 and MSK-A1. It was concluded that kaempferol had beneficial effects for the prevention of carcinogenesis.

In a study that evaluated the photoprotective activity of 4 compounds isolated from *Rhodiola crenulata*, kaempferol was shown to have the best *in vitro* antioxidant activity in the ABTS, DPPH, and FRAP tests. Since oxidative damage is a long-standing factor linked to photoaging, this compound has long been considered an effective aid to slow this process. Kaempferol exhibited significant protective effects by preventing UVR-B damage in HFS and HaCaT cell lines. This flavonoid suppressed the secretion of IL-6 and TNF-α induced by UVR-B radiation compared to other compounds [80].

Curcumin

Curcumin is a yellow-colored compound widely studied due to its known anti-inflammatory and antitumor properties. For example, Phillips *et al.*, (2011) [81] evaluated its antitumor and antimetastatic effects using squamous cell carcinoma (SCC) xenografts. The results showed that the xenografts were significantly sensitive in terms of tumor growth inhibition, even at low compound doses. Liu *et al.* (2018) [82] evaluated the effect of different doses of curcumin on HDFa cells, which were then subjected to various UVR-A intensities for two hours. Curcumin was found to modulate ROS and malonaldehyde levels, improve GSH, SOD, and CAT levels, regulate collagen metabolism by decreasing MMP-1 and MMP-3 expression, and promote the repair of cells damaged by UVR. Curcumin can be considered aviable alternative for new sunscreen formulations.

Isoflavonoids

Plant products that contain abundant isoflavonoids include soybeans, soy flour, soy milk, beer, and tempeh. Isoflavonoids have been demonstrated to be efficacious in the treatment of prostate, breast, colon, kidney, and thyroid cancer [65]. Lin in 2007 [83] evaluated the efficiency by which 0.5% isoflavone topical solutions (genistein, daidzein, biochanin A, and formononetin) in protected guinea pig cells from UV-induced sunburn erythema formation. The results obtained showed that genistein, daidzein, and biochanin A had significant photoprotective effects, while formononetin, although it did not prevent sunburn, was effective in reducing erythema.

Genistein

Genistein modulates the process of photocarcinogenesis by improving the activity of antioxidant enzymes and eliminating ROS. In a study by Cui *et al.* (2017) [84], genistein was shown to alter the morphology and migration of B16F10 murine melanoma cells, inducing apoptosis in a concentration-dependent manner. This isoflavone also prevents UV-B-induced burns in hairless mice. The antitumor activity of genistein is associated with its regulation of two pathways: preventing inflammation due to its protein tyrosine kinase activity and regulating the expression of proto-oncogenes involved in cell replacement [85].

Equol

The 4-7-isoflavondiol or equol is an isoflavone metabolite of daidzein produced in the intestinal microbiota of mammals. In model animals, the researchers topically applied equol, this isoflavone was shown to prevent the formation of CPDs and erythema. Equol has also been shown to have a protective effect on immunosuppression induced by cis-urocanic acid, indicating its potential mechanism of inactivation [86].

Anthocyanins

The anthocyanin structure is based on the 2-phenylchromenyl cation ($C_6C_3C_6$), which is biosynthetically originats in the shikimic acid pathway [87]. Anthocyanins are characterized by being highly soluble in water since they are generally found as glycosides stored in specialized vacuoles of plant cells [88]. Anthocyanins are pigments present in the flowers, roots, and fruits of many plants and show red, purple, and blue colors; in fact, their name comes from the Greek words "anthos" for flower and "kyáneos" for dark blue. The aromatic portion when not attached to a sugar is known as anthocyanidin (aglycone), among which there are 25 different forms depending on the substituents, which can be hydroxyl

(OH) or methoxy (OCH_3) groups attached to the 2-phenylchromenylium nucleus. Glycosylation is achieved enzymatically after adding the monosaccharide at the C3 and/or C5 position of the aglycone. The most common monosaccharide in anthocyanins is glucose, although galactose, rhamnose, arabinose, and xylose residues can also be present [89].

Anthocyanins have several functions in plants; for example, the color they provide to flowers and fruits has effects on pollination and seed dispersal, and they can also act as a defense mechanism against herbivores [90]. These molecules also protect against UVR-B and the intensity of other types of light radiation. Likewise, anthocyanins protect plants against oxidative stress caused by heat, water, or nutrient deficits [91].

Many of the beneficial effects of anthocyanins on human health are due to their antioxidant activity and light absorption in both the visible and UV spectra. Anthocyanins have been associated with preventing cardiovascular and neurodegenerative diseases, cancer, and diabetes [88].

In an aqueous solution, anthocyanins oscillate in equilibrium between two structural types, the flavylium cation and the quinonoidal anhydrous base, in a pH-dependent manner. The diene moiety of the quinonoidal anhydrous base is a good substrate for singlet oxygen and other ROS. Because the structures of anthocyanins can donate hydrogen atoms, it has been speculated that these compounds have antioxidant capacity *in vivo*. Anthocyanins have been shown to scavenge metabolically generated superoxide and nitric oxide radicals, decrease H_2O_2-mediated cytotoxicity, and reduce the α-tocopheroxyl radical (TocO') to a-tocopherol (TocOH). The *in vitro* antioxidant activity of anthocyanins has even been reported to supersede that of vitamin E [92]. Petruck *et al.* 2017 [93] demonstrated that *Euterpe oleracea* maldivina protects BALB/3T3 cells against UVR-A by interfering with ROS generation and maintaining normal intracellular reduced-glutathione (GSH) levels and lipid peroxidation. Pyridinium bisretinoid A2E, an autofluorescent pigment that accumulates in retinal pigment epithelial cells with age and in some retinal disorders, may mediate the detergent-like disruption of cell membranes and cellular damage induced by light. In *in vitro* experiments, it has been shown that anthocyanins can act as antioxidants that suppress photooxidative processes initiated in adult human retinal pigment epithelial (RPE) cells through the fluorophore lipofuscin A2E, reducing the damage caused by the light [92].

The protective effects of cyanidin-3-*O*-P-glucopyranoside (C-3-G) have been evaluated in UVR-A-induced apoptosis and DNA fragmentation in human keratinocytes (HaCaT cells), and treating HaCaT cells with C-3-G prior to UVR-

A irradiation inhibited apoptosis and DNA fragmentation. The antioxidant properties of C-3-G have also been investigated in HaCat cells in terms of inhibiting the formation of ROS at apoptosis-inducing doses of UVR-A, and it was found that C-3-G inhibits the release of hydrogen peroxide (H_2O_2) (an indicator of cellular ROS formation) after UV-A irradiation. Furthermore, antioxidant activity in terms of Trolox equivalents, treatment with C-3-G leads to a greater increase in antioxidant activity in the membrane-enriched fraction than in the cytosol [94]. Anthocyanins may act as stimulators of the immune response, or they may induce gene suppression or block oxidative DNA damage. Blackberry anthocyanins were shown to reduce UV-mediated oxidative injury to skin keratinocytes. Blueberry anthocyanins have been found to possess protective effects against UV-B-induced skin photoaging. This occurs by blocking collagen destruction and inflammatory responses through NF-κB transcriptional mechanisms along with MAPK signaling [95].

Tannins

Ellagic Acid

Ellagic acid is a polyphenolic compound belonging to the ellagitannin family that is present in many berries, pomegranates, grapes, and nuts. The antitumor potential of ellagic acid has been demonstrated in several experiments. Hseu *et al.* (2012) [96] demonstrated that in human keratinocytes (HaCat) pretreatment of cells with ellagic acid at concentrations of 25–75 μM, reduced UVA-induced ROS production and inhibited apoptosis. The authors also demonstrated that the antioxidant effect of ellagic acid occurred through the induction of the expression of different endogenous antioxidant enzymes, such as superoxide dismutases (SODs) and heme oxygenase 1 (HO-1), as well as the upregulation of the oxidative stress marker NF-E2-related factor-2 (Nrf2).

Tannic Acid

Gwak *et al.* (2021) [97] evaluated the UV-Vis absorbance of hyaluronic acid and tannic acid (HA/TA) hydrogel preparation. Hyaluronic acid did not show any absorbance in the UV range used, indicating that it has no UVR-blocking capacity. However, the HA/TA hydrogel showed excellent UVR absorbance over a wide range of UV radiations. The chemical structure of tannic acid contains many chromophores for UVR in addition to functional groups such as phenols and keto groups (Fig. **1**); therefore, this hydrogel could function as a barrier or screen to prevent UVR penetration into the skin. The polarity of tannic acid makes its presence in hydrogels very homogeneous, favoring topical application for sunscreen formulations, and preventing the oxidative stress and tissue damage caused by UVR.

In 2022, Daré *et al.* [98] compared the antioxidant activity of tannic acid to that of its base compound gallic acid and discovered that the antioxidant capacity of tannic acid is superior. This is because tannic acid has abundant hydroxyl groups in its structure, allowing it to neutralize free radicals. In addition, the galloyl groups present in the tannic acid molecule enhance the inhibitory effect on proteolytic enzymes because each benzene contains three hydroxyl groups that can bind with the hydrophilic groups of the extracellular matrix proteins. The many benzene rings of tannic acid, which are chromophore structures that absorb UV light, make it an excellent photoprotective candidate that prevents photoaging, extracellular matrix damage, and oxidative damage from ROS removal.

Lignins

Lignins are plant materials that naturally absorb and block UVR. Lignins can absorb UVR due to their aromatic structure and the presence of numerous phenolic, ketonic, and intramolecular hydrogen bonds (Fig. **1**). The 2-3 hydroxyl and/or methoxy substituents on the rings of technical lignins and the guaiacyl, syringyl and catechol types cause a redshift in UV absorbance, favoring UVR-B absorbance [99]. The use of lignins as a replacement for synthetic ingredients is promising in the production of sunscreens because lignins have been recognized to be less polluting and more efficient. For example, the modification of a sunscreen formulation with lignosulfonate in combination with TiO_2 showed that esterification occurred between the carboxyl groups of lignin and the hydroxyl group of TiO_2. The lignin coating not only improved the availability of TiO_2, but it also significantly increased its ability to block UVR. In addition, the SPF values of the creams containing 5, 10, and 20% lignin and TiO_2 were 16, 26, and 48, respectively [100].

Kaur *et al.* (2020) [101] demonstrated the ability of two types of lignins to be used as photoprotective agents and ZnO stabilizers. Lignin-ZnO nanocomposites incorporated as additives in neutral creams showed 5 times greater efficacy in blocking UV rays than native lignins and 2.5 times more efficacy than ZnO particles, demonstrating the additive potential of the lignin in combination with ZnO in sunscreens.

CONCLUSION

Exposure to UVR favors the production and accumulation of ROS, which contribute to the development of skin cancer. During this process, an imbalance in endogenous antioxidant systems occurs, which causes oxidative stress, inflammation, and remodeling of the extracellular matrix. The ROS act as inducers of proinflammatory genes and the production of proinflammatory mediators by cells and plasmatic mediators. When the inflammation caused by UVR becomes chronic, it can lead to a higher chance of developing skin cancer. Photocarcinogenesis is a process that involves the accumulation of CPDs, suppression of the immune system, depletion of antioxidant defenses, increased prostaglandin synthesis, and induction of ornithine decarboxylase and cyclooxygenase. Compounds with a photochemoprotective effect can prevent skin neoplasms, and when applied to the skin, they form a protective barrier against UVR Moreover, these compounds can be synthetic or natural and inorganic or organic. Phenolic compounds are natural products that can reduce the damage caused by UVR and prevent the excessive production of ROS, thereby preventing the development of skin cancer. Notable groups of phenolic compounds include flavonoids, phenylpropanoids, phenolic acids, stilbenes, coumarins, tannins, and lignins. Research in this field of knowledge is advancing and there is a view that with the discovery of new phenolic compounds obtained from plants, it will be possible to prevent and, where appropriate, treat photocarcinogenesis.

LIST OF ABBREVIATIONS

ABTS - 2	2'-Azino-bis-3-ethylbenzothiazoline-6-sulfonic acid
AKT	Serine/threonine kinase-1
AP-1	Activator protein-1
BCC	Basal cell carcinoma
CAPE	Caffeic acid phenethyl ester
CAT	Catalase
CNM	Nonmelanoma skin cancers
CPDs	Cyclobutane pyrimidine dimers
DPPH - 2	2-Diphenyl-1-picrylhydrazyl
EHMC	2-Ethylhexyl-4-methoxycinnamate
ERCC1	Excision repair cross-complementation group 1
FRAP	Ferric reducing antioxidant power
GSH	Reduced glutathione
GSH-Px	Glutathione peroxidase
HA/TA	Hyaluronic acid and tannic acid
HDF	Human dermal fibroblasts
HO-1	Heme oxygenase 1
IL-1	Interleukin-1
MAPK	Mitogen-activated protein kinases
MMPs	Matrix metalloproteinases
NER	Nucleotide excision repair

NF-kB	Nuclear factor kappa B
NO	Nitric oxide
NRF1	Nuclear respiratory factor 1
Nrf2/ARE	Nrf2/antioxidant response element
PABA	para-Aminobenzoic acid
PGC1α	Peroxisome proliferator-activated receptor γ coactivator 1 α
PI3K	Phosphatidylinositol 3 kinase
PPAR	Peroxisome proliferator-activated receptor
PPARγ	Peroxisome proliferator-activated receptor gamma
PTEN	Phosphatase and tensin homolog
RPE	Retinal pigment epithelial
SCC	Squamous cell carcinoma
SOD	Superoxide dismutase
SPF	Sun protection factor
STAT-3	Signal transducer and activator of transcription 3
TGF-β	Transforming growth factor beta
TIFIIH	Transcription factor II H
TNF-α	Tumor necrosis factor-α
TocO'	α-Tocopheroxyl radical
TocOH	α-Tocopherol
XPC	Xeroderma pigmentosum complementation group C
XPE	Xeroderma pigmentosum complementation group E

REFERENCES

[1] Arnold M, Singh D, Laversanne M, *et al.* Global burden of cutaneous melanoma in 2020 and projections to 2040. JAMA Dermatol 2022; 158(5): 495-503.
[http://dx.doi.org/10.1001/jamadermatol.2022.0160] [PMID: 35353115]

[2] International Agency for Research on Cancer. IARC working group on the evaluation of carcinogenic risk to humans. 2012.

[3] Gill SS, Anjum NA, Gill R, Jha M, Tuteja N. DNA damage and repair in plants under ultraviolet and ionizing radiations. ScientWorldJ 2015; 2015(250158): 1-11.
[http://dx.doi.org/10.1155/2015/250158] [PMID: 25729769]

[4] Sharma S, Chatterjee S, Kataria S, *et al.* Review on responses of plants to UV-B radiation related stress.UV-B Radiation. USA: John Wiley & Sons Inc. 2017; pp. 75-97.
[http://dx.doi.org/10.1002/9781119143611.ch5]

[5] Gefeller O, Diehl K. Children and ultraviolet radiation. Children 2022; 9(4): 537.
[http://dx.doi.org/10.3390/children9040537] [PMID: 35455581]

[6] Rhodes LE, Belgi G, Parslew R, McLoughlin L, Clough GF, Friedmann PS. Ultraviolet-B-induced erythema is mediated by nitric oxide and prostaglandin E2 in combination. J Invest Dermatol 2001;

117(4): 880-5.
[http://dx.doi.org/10.1046/j.0022-202x.2001.01514.x] [PMID: 11676827]

[7]   Phillips JM, Clark C, Herman-Ferdinandez L, *et al.* Curcumin inhibits skin squamous cell carcinoma tumor growth *in vivo.* Otolaryngol Head Neck Surg 2011; 145(1): 58-63.
[http://dx.doi.org/10.1177/0194599811400711] [PMID: 21493306]

[8]   Terra VA, Souza-Neto FP, Pereira RC, *et al.* Time-dependent reactive species formation and oxidative stress damage in the skin after UVB irradiation. J Photochem Photobiol B 2012; 109: 34-41.
[http://dx.doi.org/10.1016/j.jphotobiol.2012.01.003] [PMID: 22356772]

[9]   McAdam E, Brem R, Karran P. Oxidative stress–induced protein damage inhibits DNA repair and determines mutation risk and therapeutic efficacy. Mol Cancer Res 2016; 14(7): 612-22.
[http://dx.doi.org/10.1158/1541-7786.MCR-16-0053] [PMID: 27106867]

[10]   Kochevar IE, Taylor CR, Krutmann J. Fundamentos de fotobiología y fotoinmunología cutáneas. Dermatología en Medicina General Tomo I 8ª Edición. España: Editorial Medica Panamericana 2014; pp. 55-8.

[11]   Nichols JA, Katiyar SK. Skin photoprotection by natural polyphenols: anti-inflammatory, antioxidant and DNA repair mechanisms. Arch Dermatol Res 2010; 302(2): 71-83.
[http://dx.doi.org/10.1007/s00403-009-1001-3] [PMID: 19898857]

[12]   Kindt TJ, Goldsby RA, Osborne BA. Kuby Immunology. 6th. England: W. H. Freeman and Company 2007; p. 57.

[13]   Gloire G, Legrand-Poels S, Piette J. NF-κB activation by reactive oxygen species: Fifteen years later. Biochem Pharmacol 2006; 72(11): 1493-505.
[http://dx.doi.org/10.1016/j.bcp.2006.04.011] [PMID: 16723122]

[14]   Liu T, Zhang L, Joo D, Sun SC. NF-κB signaling in inflammation. Signal Transduct Target Ther 2017; 2(1): 17023.
[http://dx.doi.org/10.1038/sigtrans.2017.23] [PMID: 29158945]

[15]   Zheng C, Yin Q, Wu H. Structural studies of NF-κB signaling. Cell Res 2011; 21(1): 183-95.
[http://dx.doi.org/10.1038/cr.2010.171] [PMID: 21135870]

[16]   Yaar M, Gilchrest BA. Envejecimiento de la piel. 2014.

[17]   Thiers BH, Maize JC, Spicer SS, Cantor AB. The effect of aging and chronic sun exposure on human Langerhans cell populations. J Invest Dermatol 1984; 82(3): 223-6.
[http://dx.doi.org/10.1111/1523-1747.ep12260055] [PMID: 6199432]

[18]   Tobin DJ. Introduction to skin aging. J Tissue Viability 2017; 26(1): 37-46.
[http://dx.doi.org/10.1016/j.jtv.2016.03.002] [PMID: 27020864]

[19]   López-Camarillo C, Aréchaga Ocampo E, López Casamichana M, Pérez-Plasencia C, Álvarez-Sánchez E, Marchat LA. Protein kinases and transcription factors activation in response to UV-radiation of skin: Implications for carcinogenesis. Int J Mol Sci 2011; 13(1): 142-72.
[http://dx.doi.org/10.3390/ijms13010142] [PMID: 22312244]

[20]   Kashyap MP, Sinha R, Mukhtar MS, Athar M. Epigenetic regulation in the pathogenesis of non-melanoma skin cancer. Semin Cancer Biol 2022; 83(83): 36-56.
[http://dx.doi.org/10.1016/j.semcancer.2020.11.009] [PMID: 33242578]

[21]   Jones DL, Baxter BK. DNA repair and photoprotection: Mechanisms of overcoming environmental ultraviolet radiation exposure in halophilic archaea. Front Microbiol 1882; 2017(8): 1-16.
[PMID: 29033920]

[22]   Garg C, Sharma H, Garg M. Skin photo-protection with phytochemicals against photo-oxidative stress, photo-carcinogenesis, signal transduction pathways and extracellular matrix remodeling: An overview. Ageing Res Rev 2020; 62(101127): 101127.
[http://dx.doi.org/10.1016/j.arr.2020.101127] [PMID: 32721499]

[23] Afaq F. Photocarcinogenesis. Encyclopedia of Cancer. Heidelberg: Springer Berlin 2011; pp. 2870-4.
      [http://dx.doi.org/10.1007/978-3-642-16483-5_4546]

[24] Matsumura Y, Ananthaswamy HN. Toxic effects of ultraviolet radiation on the skin. Toxicol Appl
      Pharmacol 2004; 195(3): 298-308.
      [http://dx.doi.org/10.1016/j.taap.2003.08.019] [PMID: 15020192]

[25] Baliga MS, Katiyar SK. Chemoprevention of photocarcinogenesis by selected dietary botanicals.
      Photochem Photobiol Sci 2006; 5(2): 243-53.
      [http://dx.doi.org/10.1039/b505311k] [PMID: 16465310]

[26] Linares MA, Zakaria A, Nizran P. Skin Cancer. Prim Care 2015; 42(4): 645-59.
      [http://dx.doi.org/10.1016/j.pop.2015.07.006] [PMID: 26612377]

[27] Donaldson MJ, Sullivan TJ, Whitehead KJ, Williamson RM. Squamous cell carcinoma of the eyelids.
      Br J Ophthalmol 2002; 86(10): 1161-5.
      [http://dx.doi.org/10.1136/bjo.86.10.1161] [PMID: 12234899]

[28] Seidl-Philipp M, Nguyen VA. Cutaneous melanoma-The benefit of screening and preventive
      measures. memo 2019; 12: 235-8.

[29] Mota MD, Costa RYS, Guedes AS, Silva LCRC, Chinalia FA. Guava-fruit extract can improve the
      UV-protection efficiency of synthetic filters in sun cream formulations. J Photochem Photobiol B
      2019; 201: 111639.
      [http://dx.doi.org/10.1016/j.jphotobiol.2019.111639] [PMID: 31698220]

[30] Napagoda MT, Malkanthi BMAS, Abayawardana SAK, Qader MM, Jayasinghe L. Photoprotective
      potential in some medicinal plants used to treat skin diseases in Sri Lanka. BMC Complement Altern
      Med 2016; 16(1): 479-85.
      [http://dx.doi.org/10.1186/s12906-016-1455-8] [PMID: 27881112]

[31] Catelan TBS, Gaiola L, Duarte BF, Cardoso CAL. Evaluation of the *in vitro* photoprotective potential
      of ethanolic extracts of four species of the genus *Campomanesia*. J Photochem Photobiol B 2019; 197:
      111500.
      [http://dx.doi.org/10.1016/j.jphotobiol.2019.04.009] [PMID: 31200215]

[32] Rezende SG, Dourado JG, Amorim De Lino FM, Vinhal DC, Silva EC, Gil EDS. Methods used in
      evaluation of the sun protection factor of sunscreens. Revista Eletrônica de Farmácia 2014; 11(2): 37-
      54.
      [http://dx.doi.org/10.5216/ref.v11i2.27013]

[33] Sales-Barbosa J, Gonçalves-Araújo T. Lipid and polymeric nanoparticles for organic sunscreen filters.
      Rev Cuba Farm 2021; 53(4): 456-65.

[34] Kulkarni SS, Bhalke RS, Pande VV, Prakash N, Kendre PN. Herbal Plants. in photo protection and
      sun screening action: An overview, Indo Am. J Pharm Res 2014; 4(2): 1104-8.

[35] Pico Y, Belenguer V, Corcellas C, *et al.* Contaminants of emerging concern in freshwater fish from
      four Spanish Rivers. Sci Total Environ 2019; 659: 1186-98.
      [http://dx.doi.org/10.1016/j.scitotenv.2018.12.366] [PMID: 31096332]

[36] Penta D, Somashekar BS, Meeran SM. Epigenetics of skin cancer: Interventions by selected bioactive
      phytochemicals. Photodermatol Photoimmunol Photomed 2018; 34(1): 42-9.
      [http://dx.doi.org/10.1111/phpp.12353] [PMID: 28976029]

[37] Iqbal J, Abbasi BA, Ahmad R, *et al.* Potential phytochemicals in the fight against skin cancer: Current
      landscape and future perspectives. Biomed Pharmacother 2019; 109: 1381-93.
      [http://dx.doi.org/10.1016/j.biopha.2018.10.107] [PMID: 30551389]

[38] Muhammad A, Feng X, Rasool A, Sun W, Li C. Production of plant natural products through
      engineered *Yarrowia lipolytica*. Biotechnol Adv 2020; 43(43): 107555.
      [http://dx.doi.org/10.1016/j.biotechadv.2020.107555] [PMID: 32422161]

[39]   Misra D, Dutta W, Jha G, Ray P. Interactions and regulatory functions of phenolics in soil-plant-climate nexus. Agronomy 2023; 13(2): 280.
[http://dx.doi.org/10.3390/agronomy13020280]

[40]   Liu X, Zhang R, Shi H, *et al.* Protective effect of curcumin against ultraviolet A irradiation-induced photoaging in human dermal fibroblasts. Mol Med Rep 2018; 17(5): 7227-37.
[http://dx.doi.org/10.3892/mmr.2018.8791] [PMID: 29568864]

[41]   Gado AR, Ellakany HF, Elbestawy AR, *et al.* Herbal medicine additives as powerful agents to control and prevent avian influenza virus in poultry : A review. Ann Anim Sci 2019; 19(4): 905-35.
[http://dx.doi.org/10.2478/aoas-2019-0043]

[42]   Briguglio G, Costa C, Pollicino M, Giambò F, Catania S, Fenga C. Polyphenols in cancer prevention: New insights (Review). Int J Funct Nutr 2020; 1(2): 9-20.
[http://dx.doi.org/10.3892/ijfn.2020.9]

[43]   Balupillai A, Nagarajan RP, Ramasamy K, Govindasamy K, Muthusamy G. Caffeic acid prevents UVB radiation induced photocarcinogenesis through regulation of PTEN signaling in human dermal fibroblasts and mouse skin. Toxicol Appl Pharmacol 2018; 352: 87-96.
[http://dx.doi.org/10.1016/j.taap.2018.05.030] [PMID: 29802912]

[44]   Pluemsamran T, Onkoksoong T, Panich U. Caffeic acid and ferulic acid inhibit UVA-induced matrix metalloproteinase-1 through regulation of antioxidant defense system in keratinocyte HaCaT cells. Photochem Photobiol 2012; 88(4): 961-8.
[http://dx.doi.org/10.1111/j.1751-1097.2012.01118.x] [PMID: 22360712]

[45]   Balupillai A, Prasad RN, Ramasamy K, *et al.* Caffeic acid inhibits UVB-induced inflammation and photocarcinogenesis through activation of peroxisome proliferator-activated receptor-γ in mouse skin. Photochem Photobiol 2015; 91(6): 1458-68.
[http://dx.doi.org/10.1111/php.12522] [PMID: 26303058]

[46]   Agilan B, Rajendra Prasad N, Kanimozhi G, *et al.* Caffeic acid inhibits chronic UVB-induced cellular proliferation through JAK- STAT-3 Signaling in Mouse Skin. Photochem Photobiol 2016; 92(3): 467-74.
[http://dx.doi.org/10.1111/php.12588] [PMID: 27029485]

[47]   Choromanska A, Saczko J, Kulbacka J. Caffeic acid phenethyl ester assisted by reversible electroporation *In Vitro* study on human melanoma cells. Pharmaceutics 2020; 12(5): 478.
[http://dx.doi.org/10.3390/pharmaceutics12050478] [PMID: 32456290]

[48]   Gopala Krishna AG, Prasanth Kumar PK. Physico chemical characteristics and nutraceutical distribution of crude palm oil and its fractions. Grasas Aceites 2014; 65(2): e018.
[http://dx.doi.org/10.3989/gya.097413]

[49]   Hwang E, Park SY, Lee HJ, Lee TY, Sun Z, Yi TH. Gallic acid regulates skin photoaging in UVB-exposed fibroblast and hairless mice. Phytother Res 2014; 28(12): 1778-88.
[http://dx.doi.org/10.1002/ptr.5198] [PMID: 25131997]

[50]   Espinosa-Alonso LG, Valdez-Morales M, Aparicio-Fernandez X, Medina-Godoy S, Guevara-Lara F. Vegetables by products. In Campos-Vega, Oomah BD, Vergara-Castañeda A, Eds. Vegetable By-products. Food Wastes and By-products: Nutraceutical and Health Potential, USA: John Wiley & Sons 2020; pp. 223-266.
[http://dx.doi.org/10.1002/9781119534167.ch8]

[51]   Zhou MW, Jiang RH, Kim KD, *et al.* Rosmarinic acid inhibits poly(I:C)-induced inflammatory reaction of epidermal keratinocytes. Life Sci 2016; 155: 189-94.
[http://dx.doi.org/10.1016/j.lfs.2016.05.023] [PMID: 27210890]

[52]   Alagawany M, Abd El-Hack ME, Farag MR, *et al.* Rosmarinic acid: Modes of action, medicinal values and health benefits. Anim Health Res Rev 2017; 18(2): 167-76.
[http://dx.doi.org/10.1017/S1466252317000081] [PMID: 29110743]

[53] Seeram NP, Kulkarni VV, Padhye S. Sources and Chemistry of Resveratrol.Resveratrol in Health and Disease. Los Angeles: CRC Press Taylor & Francis Group 2006; pp. 17-32.

[54] Wen S, Zhang J, Yang B, Elias PM, Man MQ. Role of resveratrol in regulating cutaneous functions. Evid Based Complement Alternat Med 2020; 2020: 1-20.
[http://dx.doi.org/10.1155/2020/2416837] [PMID: 32382280]

[55] Vervandier-Fasseur D, Latruffe N. The potential use of resveratrol for cancer prevention. Molecules 2019; 24(24): 4506.
[http://dx.doi.org/10.3390/molecules24244506] [PMID: 31835371]

[56] Afaq F, Adhami VM, Ahmad N, Mukhtar H. Botanical antioxidants for chemoprevention of photocarcinogenesis. Front Biosci 2002; 7(1-3): d784-92.
[http://dx.doi.org/10.2741/afaq] [PMID: 11897547]

[57] Saewan N, Jimtaisong A. Natural products as photoprotection. J Cosmet Dermatol 2015; 14(1): 47-63.
[http://dx.doi.org/10.1111/jocd.12123] [PMID: 25582033]

[58] Aziz MH, Reagan-Shaw S, Wu J, Longley BJ, Ahmad N. Chemoprevention of skin cancer by grape constituent resveratrol: Relevance to human disease? FASEB J 2005; 19(9): 1193-5.
[http://dx.doi.org/10.1096/fj.04-3582fje] [PMID: 15837718]

[59] Afaq F, Adhami VM, Ahmad N. Prevention of short-term ultraviolet B radiation-mediated damages by resveratrol in SKH-1 hairless mice part of this work was conducted at the department of dermatology, case western reserve university and the research institute of university hospitals of Cleveland, 11100 Euclid Avenue, Cleveland, Ohio 44106. Toxicol Appl Pharmacol 2003; 186(1): 28-37.
[http://dx.doi.org/10.1016/S0041-008X(02)00014-5] [PMID: 12583990]

[60] Maruki-Uchida H, Kurita I, Sugiyama K, Sai M, Maeda K, Ito T. The protective effects of piceatannol from passion fruit (*Passiflora edulis*) seeds in UVB-irradiated keratinocytes. Biol Pharm Bull 2013; 36(5): 845-9.
[http://dx.doi.org/10.1248/bpb.b12-00708] [PMID: 23649341]

[61] Du M, Zhang Z, Gao T. Piceatannol induced apoptosis through up-regulation of microRNA-181a in melanoma cells. Biol Res 2017; 50(1): 36.
[http://dx.doi.org/10.1186/s40659-017-0141-8] [PMID: 29041990]

[62] Shelan Nagapan T, Rohi Ghazali A, Fredalina Basri D, Nallance Lim W. Photoprotective effect of stilbenes and its derivatives against ultraviolet radiation-induced skin disorders. Biomed Pharmacol J 2018; 11(3): 1199-208.
[http://dx.doi.org/10.13005/bpj/1481]

[63] Brodowska K. Natural flavonoids: Classification, potential role, and application of flavonoid analogs. Eur J Biol Res 2023; 7(2): 108-23.

[64] Tuli HS, Garg VK, Bhushan S, *et al*. Natural flavonoids exhibit potent anticancer activity by targeting microRNAs in cancer: A signature step hinting towards clinical perfection. Transl Oncol 2023; 27: 101596.
[http://dx.doi.org/10.1016/j.tranon.2022.101596] [PMID: 36473401]

[65] Batra P, Sharma AK. Anti-cancer potential of flavonoids: recent trends and future perspectives. 3 Biotech 2013; 3(6): 439-59.
[http://dx.doi.org/10.1007/s13205-013-0117-5] [PMID: 28324424]

[66] Heinrich U, Neukam K, Tronnier H, Sies H, Stahl W. Long-term ingestion of high flavanol cocoa provides photoprotection against UV-induced erythema and improves skin condition in women. J Nutr 2006; 136(6): 1565-9.
[http://dx.doi.org/10.1093/jn/136.6.1565] [PMID: 16702322]

[67] Jiminez V, Yusuf N. An update on clinical trials for chemoprevention of human skin cancer. J Cancer Metastasis Treat 2023; 9(1): 4.
[PMID: 37786882]

[68]   Shankar E, Goel A, Gupta K, Gupta S. Plant flavone apigenin: An emerging anticancer agent. Curr Pharmacol Rep 2017; 3(6): 423-46.
[http://dx.doi.org/10.1007/s40495-017-0113-2] [PMID: 29399439]

[69]   Ganai SA, Sheikh FA, Baba ZA, Mir MA, Mantoo MA, Yatoo MA. Anticancer activity of the plant flavonoid luteolin against preclinical models of various cancers and insights on different signalling mechanisms modulated. Phytother Res 2021; 35(7): 3509-32.
[http://dx.doi.org/10.1002/ptr.7044] [PMID: 33580629]

[70]   Punia Bangar S, Kajla P, Chaudhary V, Sharma N, Ozogul F. Luteolin: A flavone with myriads of bioactivities and food applications. Food Biosci 2023; 52: 102366.
[http://dx.doi.org/10.1016/j.fbio.2023.102366]

[71]   Sun Y, Guo T, Sui Y, Li F. Quantitative determination of rutin, quercetin, and adenosine in *Flos Carthami* by capillary electrophoresis. J Sep Sci 2003; 26(12-13): 1203-6.
[http://dx.doi.org/10.1002/jssc.200301437]

[72]   Choquenet B, Couteau C, Paparis E, Coiffard LJM. Quercetin and rutin as potential sunscreen agents: determination of efficacy by an *in vitro* method. J Nat Prod 2008; 71(6): 1117-8.
[http://dx.doi.org/10.1021/np7007297] [PMID: 18512988]

[73]   Michaud-Levesque J, Bousquet-Gagnon N, Béliveau R. Quercetin abrogates IL-6/STAT3 signaling and inhibits glioblastoma cell line growth and migration. Exp Cell Res 2012; 318(8): 925-35.
[http://dx.doi.org/10.1016/j.yexcr.2012.02.017] [PMID: 22394507]

[74]   Moon SK, Cho GO, Jung SY, *et al.* Quercetin exerts multiple inhibitory effects on vascular smooth muscle cells: Role of ERK1/2, cell-cycle regulation, and matrix metalloproteinase-9. Biochem Biophys Res Commun 2003; 301(4): 1069-78.
[http://dx.doi.org/10.1016/S0006-291X(03)00091-3] [PMID: 12589822]

[75]   Masuda A, Maeno K, Nakagawa T, Saito H, Takahashi T. Association between mitotic spindle checkpoint impairment and susceptibility to the induction of apoptosis by anti-microtubule agents in human lung cancers. Am J Pathol 2003; 163(3): 1109-16.
[http://dx.doi.org/10.1016/S0002-9440(10)63470-0] [PMID: 12937152]

[76]   Soll F, Ternent C, Berry IM, Kumari D, Moore TC. Quercetin inhibits proliferation and induces apoptosis of B16 melanoma cells *in vitro*. Assay Drug Dev Technol 2020; 18(6): 261-8.
[http://dx.doi.org/10.1089/adt.2020.993] [PMID: 32799543]

[77]   Gęgotek A, Jarocka-Karpowicz I, Skrzydlewska E. Cytoprotective effect of ascorbic acid and rutin against oxidative changes in the proteome of skin fibroblasts cultured in a three-dimensional system. Nutrients 2020; 12(4): 1074.
[http://dx.doi.org/10.3390/nu12041074] [PMID: 32294980]

[78]   Muhammad KJ, Jamil S, Basar N. Phytochemical study and biological activities of *Scurrula parasitica* L (Loranthaceae) leaves. J Res Pharm 2019; 23(3): 522-31.
[http://dx.doi.org/10.12991/jrp.2019.159]

[79]   Yao K, Chen H, Lee MH, *et al.* Abstract 1241: Kaempferol suppresses solar ultraviolet radiation-induced skin cancers by targeting RSK2 and MSK1. Cancer Res 2014; 74(19_Supplement): 1241-1.
[http://dx.doi.org/10.1158/1538-7445.AM2014-1241] [PMID: 25136067]

[80]   Wang T, Wu Q, Zhao T. Preventive effects of kaempferol on high-fat diet-induced obesity complications in C57BL/6 mice. BioMed Res Int 2020; 2020: 1-9.
[http://dx.doi.org/10.1155/2020/4532482] [PMID: 32337249]

[81]   Wilken R, Veena MS, Wang MB, *et al.* Curcumin: A review of anti-cancer properties and therapeutic activity in head and neck squamous cell carcinoma. Mol Cancer 2011; 10(12)
[http://dx.doi.org/https://doi.org/10.1186/1476-4598-10-12 81]

[82]   Liu X, Zhang R, Shi H, *et al.* Protective effect of curcumin against ultraviolet A irradiation induced photoaging in human dermal fibroblasts. Mol Med Rep 2018; 17(5): 7227-37.

[http://dx.doi.org/10.3892/mmr.2018.8791] [PMID: 29568864]

[83] Lin JY, Tournas JA, Burch JA, Monteiro-Riviere NA, Zielinski J. Topical isoflavones provide effective photoprotection to skin. Photodermatol Photoimmunol Photomed 2008; 24(2): 61-6.
[http://dx.doi.org/10.1111/j.1600-0781.2008.00329.x] [PMID: 18353084]

[84] Cui S, Wang J, Wu Q, *et al.* Genistein inhibits the growth and regulates the migration and invasion abilities of melanoma cells *via* the FAK/paxillin and MAPK pathways. Oncotarget 2017; 8(13): 21674-91.
[http://dx.doi.org/10.18632/oncotarget.15535] [PMID: 28423510]

[85] Wei H, Saladi R, Lu Y, *et al.* Isoflavone genistein: Photoprotection and clinical implications in dermatology. J Nutr 2003; 133(11) (1): 3811S-9S.
[http://dx.doi.org/10.1093/jn/133.11.3811S] [PMID: 14608119]

[86] Saewan N, Jimtaisong A. Photoprotection of natural flavonoids. J Appl Pharm Sci 2013; 3(9): 129-41.

[87] Winkel-Shirley B. Molecular genetics and control of anthocyanin expression. Adv Bot Res 2002; 37: 75-94.
[http://dx.doi.org/10.1016/S0065-2296(02)37044-7]

[88] Câmara JS, Locatelli M, Pereira JAM, *et al.* Behind the scenes of anthocyanins-from the health benefits to potential applications in food, pharmaceutical and cosmetic fields. Nutrients 2022; 14(23): 5133.
[http://dx.doi.org/10.3390/nu14235133] [PMID: 36501163]

[89] Andersen ØM, Jordheim M. Basic anthocyanin chemistry and dietary sources.Anthocyanins in health and disease. Los Angeles: CRC Press Taylor & Francis Group 2013; pp. 13-89.

[90] Schaefer HM. Why fruits go to the dark side. Acta Oecol 2011; 37(6): 604-10.
[http://dx.doi.org/10.1016/j.actao.2011.04.008]

[91] Merzlyak MN, Chivkunova OB. Light-stress-induced pigment changes and evidence for anthocyanin photoprotection in apples. J Photochem Photobiol B 2000; 55(2-3): 155-63.
[http://dx.doi.org/10.1016/S1011-1344(00)00042-7] [PMID: 10942080]

[92] Jang YP, Zhou J, Nakanishi K, Sparrow JR. Anthocyanins protect against A2E photooxidation and membrane permeabilization in retinal pigment epithelial cells. Photochem Photobiol 2005; 81(3): 529-36.
[http://dx.doi.org/10.1111/j.1751-1097.2005.tb00221.x] [PMID: 15745429]

[93] Petruk G, Illiano A, Del Giudice R, *et al.* Malvidin and cyanidin derivatives from açai fruit (*Euterpe oleracea* Mart.) counteract UV-A-induced oxidative stress in immortalized fibroblasts. J Photochem Photobiol B 2017; 172: 42-51.
[http://dx.doi.org/10.1016/j.jphotobiol.2017.05.013] [PMID: 28527426]

[94] Tarozzi A, Marchesi A, Hrelia S, *et al.* Protective effects of cyanidin-3-O-β-glucopyranoside against UVA-induced oxidative stress in human keratinocytes. Photochem Photobiol 2005; 81(3): 623-9.
[PMID: 15701043]

[95] Diaconeasa Z, Ştirbu I, Xiao J, *et al.* Anthocyanins, vibrant color pigments, and their role in skin cancer prevention. Biomedicines 2020; 8(9): 336.
[http://dx.doi.org/10.3390/biomedicines8090336] [PMID: 32916849]

[96] Hseu YC, Chou CW, Senthil Kumar KJ, *et al.* Ellagic acid protects human keratinocyte (HaCaT) cells against UVA-induced oxidative stress and apoptosis through the upregulation of the HO-1 and Nrf-2 antioxidant genes. Food Chem Toxicol 2012; 50(5): 1245-55.
[http://dx.doi.org/10.1016/j.fct.2012.02.020] [PMID: 22386815]

[97] Gwak MA, Hong BM, Park WH. Hyaluronic acid/tannic acid hydrogel sunscreen with excellent anti-UV, antioxidant, and cooling effects. Int J Biol Macromol 2021; 191: 918-24.
[http://dx.doi.org/10.1016/j.ijbiomac.2021.09.169] [PMID: 34597695]

[98]   Daré RG, Nakamura CV, Ximenes VF, Lautenschlager SOS. Tannic acid, a promising anti-photoaging agent: Evidences of its antioxidant and anti-wrinkle potentials, and its ability to prevent photodamage and MMP-1 expression in L929 fibroblasts exposed to UVB. Free Radic Biol Med 2020; 160: 342-55.
[http://dx.doi.org/10.1016/j.freeradbiomed.2020.08.019] [PMID: 32858160]

[99]   Widsten P. Lignin based sunscreens state of the art, prospects and challenges. Cosmetics 2020; 7(4): 85-93.
[http://dx.doi.org/10.3390/cosmetics7040085]

[100]   Yu J, Li L, Qian Y, Lou H, Yang D, Qiu X. Facile and green preparation of high UV-blocking lignin/titanium dioxide nanocomposites for developing natural sunscreens. Ind Eng Chem Res 2018; 57(46): 15740-8.
[http://dx.doi.org/10.1021/acs.iecr.8b04101]

[101]   Kaur R, Thakur NS, Chandna S, Bhaumik J. Development of agri-biomass based lignin derived zinc oxide nanocomposites as promising UV protectant-cum-antimicrobial agents. J Mater Chem B Mater Biol Med 2020; 8(2): 260-9.
[http://dx.doi.org/10.1039/C9TB01569H] [PMID: 31799593]

CHAPTER 6

Natural Products in Wound Regeneration

Nallely Álvarez-Santos[1,2]**, Rocío Serrano-Parrales**[3]**, Patricia Guevara-Fefer**[4]**, Felix Krengel**[4] **and Ana María García-Bores**[1,*]

[1] *Phytochemistry Laboratory, UBIPRO, Superior Studies Faculty (FES)-Iztacala, National Autonomous University of Mexico (UNAM), Tlalnepantla de Baz, Mexico State, 54090, Mexico*

[2] *Postgraduate Biological Sciences, Postgraduate Studies Unit, National Autonomous University of Mexico (UNAM), Coyoacan, Mexico City, 04510, Mexico*

[3] *Laboratory of Bioactivity of Natural Products, UBIPRO, Superior Studies Faculty (FES)-Iztacala, National Autonomous University of Mexico (UNAM), Tlalnepantla de Baz, México State, 54090, México*

[4] *Department of Ecology and Natural Products, Faculty of Sciences, National Autonomous University of Mexico (UNAM), Coyoacan, Mexico City, 04510, Mexico*

Abstract: The skin is the largest organ in the body that provides protection. When a wound occurs, the skin structure and its function are damaged, and it can even compromise life. Damage repair can occur through two mechanisms: healing and regeneration. When a scar forms, fibrosis occurs in the area, and the skin appendages, which include the glands and hair follicles, are lost. In regeneration, the functionality of the skin is partially or totally recovered. Medicinal plants and their active principles favor the regeneration of skin wounds because they have direct effects on the different phases of the process. They favor hemostasis, and modulate inflammation, which allows the following stages of healing to occur in less time, such as proliferation and remodeling. They favor hemostasis, modulate inflammation, and that the following stages of healing to occur in less time (proliferation and remodeling). Natural products can also reduce the risk of wound infections by having antibacterial activity. However, the bioavailability of the extracts and their metabolites may be limited, and a solution to this problem is to integrate them into preparations such as hydrogels, nanoparticles, nanofibers, and nanoemulsions. Research on the therapeutic properties of various natural products and their integration into the formulations mentioned above for wound regeneration is described below according to their effect on epithelialization, regeneration of epidermal appendages, vascularization, and in some cases their mechanism of action.

Keywords: Hair follicles, Hydrogels, Nanoemulsions, Nanofibers, Nanoparticles, Natural products, Wound regeneration.

---

* **Corresponding author Ana María García-Bores:** Phytochemistry Laboratory, UBIPRO, Superior Studies Faculty (FES)-Iztacala, National Autonomous University of Mexico (UNAM), Tlalnepantla de Baz, Mexico State, 54090, Mexico; Tel: +52555556231136; E-mail: boresana@iztacala.unam.mx

Israel Valencia Quiroz (Ed.)

INTRODUCTION

The largest organ in the body is the skin; one of its main functions is to maintain the integrity of the individual. It has two main tissues: a) the outermost is the epidermis, a keratinized squamous epithelium composed mainly of keratinocytes, and the thickness varies from 0.04 to 1.6 mm depending on the location; b) the dermis is made up mainly of fibroblasts that synthesize and remodel the extracellular matrix (ECM) such as type I collagen fibers, elastic fibers, and ground substance; this layer is thicker than the epidermis -15 to 40 times-epidermis [1, 2].

In the dermis, there are two important compartments: a) the papillary dermis (PD) which is located below the epidermis, and is rich in papillary fibroblasts, with a high proliferative and synthetic capacity; and b) the reticular dermis (RD) which is much thicker with a parallel organization of connective tissue fibers to the surface of the skin, below the fatty layer called the hypodermis and later the fascia. In these layers, we find the radicular fibroblasts that oversee the synthesis of the ECM and participate in the generation of adipocytes. The epidermis and dermis are the hair follicle (HF) associated with sebaceous glands and the erector pili muscle, known as epidermal appendages [1]. It has been reported that papillary fibroblasts participate in the regulation of the maintenance and growth of the epidermis, as well as of the follicles. Mesenchymal stem cells are found in PD located at the base of the HF [2].

When the integrity of the skin is altered, a wound occurs. The damage repair process is carried out through the formation of scars or fibrotic repair that do not present HF, sebaceous glands or sweat glands. The wound healing process comprises four interposed phases: a) hemostasis (generation of a clot), b) inflammation (debridement of the wound), c) proliferation (activation of fibroblasts with collagen III synthesis), and d) remodeling (restoring type I collagen), resulting in a scar [2].

The process of healing wounds on the skin occurs in several stages and there are no specific time limits between them, they overlap each other. This process is dynamic and highly regulated by cellular, humoral, and molecular mechanisms that participate in each of the phases. It begins immediately after the injury and can last for years. Closure of skin wounds can be achieved through scarring or regeneration. Healing occurs through a nonspecific form of fibrosis and scar formation [3].

On the other hand, skin regeneration consists of the replacement and specific proliferation of tissues, such as the epidermis, and dermis with their annexes such as HFs [3, 4]. The role of papillary fibroblasts is crucial in skin regeneration by preventing fibrotic effects, and promoting the development of the epidermis, and neoformation of HFs and blood vessels [2]. After an injury, PD fibroblasts respond to Wnt/β-catenin and Sonic Hedgehog (Shh) signals. Transforming epidermal growth factor beta (TGF-β) stimulates DR fibroblasts to proliferate and secrete ECM in regeneration [5]. The epidermal Shh signaling pathway is involved in reseating dermal papillae with the regenerative niche that promotes hair follicle neogenesis (HFN). The involvement of different pathways is complex and can lead to scarring or regeneration; for example, sustained Wnt expression is associated with fibrosis, but Shh signaling in Wnt-active cells promotes dermal papillae [6]. The temporal and spatial regulation of signals that can induce fibrotic scar formation or regeneration is complex and continues to be the subject of extensive research (Fig. **1**).

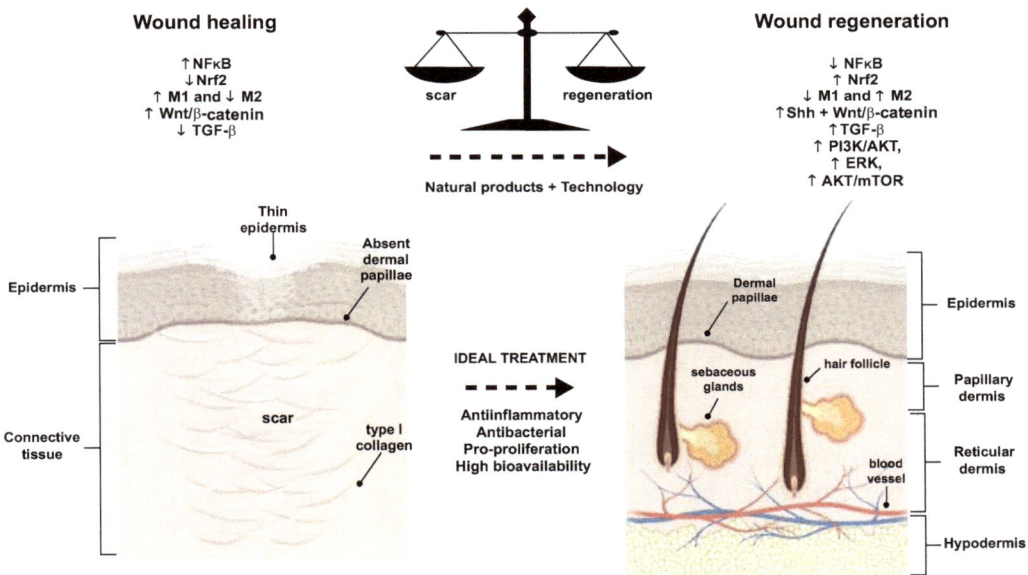

Fig. (1). Differences between wound healing and wound regeneration. ↑: increase; ↓: decrease, M1: type 1 macrophage, M2: type 2 macrophage. Created in BioRender.

NATURAL PRODUCTS: EXTRACTS AND SECONDARY METABOLITES

Treatments for wound healing are diverse and with them, the aim is to promote the healing phases and regeneration of tissue. In addition, it is recommended that treatments help maintain humidity around the wound, promote gas exchange, prevent infection, and be biocompatible, biodegradable, and non-toxic. In this sense, natural products have multiple mechanisms that can affect the different phases of healing, and can even induce regeneration by promoting re-epithelialization, de novo formation of HFs, and vascularization [7].

The extracts of several plant species, as well as compounds derived from their secondary metabolism, have been used for the treatment of wounds. Some have an effect in the early phases of tissue repair, favoring hemostasis, and reducing inflammation time, which allows the following stages of healing to occur in less time, such as proliferation and remodeling. Other compounds show an indirect beneficial effect; for example, there are active ingredients that have antioxidant, antimicrobial, or anti-inflammatory activity [7].

Chronic diseases decrease the healing process and produce chronic wounds [8]. Diabetic patients with wounds are more susceptible to infection, which can cause amputation of members, bacteremia, and death [9]. Researchers are focusing on treatment in which wounds are closed quickly, infection is prevented, and tissue regeneration is promoted. One strategy is the application and use of technology tools.

The antimicrobial activity of an active substance favors healing by inhibiting the development of microorganisms, which prevents wounds from becoming infected so that the tissue repair process can continue its course and culminate in an adequate time. As previously mentioned, inflammation is a stage of healing, and it must occur since growth factors are produced that allow the proliferation of keratinocytes and fibroblasts, as well as the production of collagen, ECM, and re-epithelialization. Growth factors involved in the formation of blood vessels, hair follicles, and sebaceous glands are also produced [7]. Another important point is the change in the phenotype of macrophages. M1 macrophages are a proinflammatory phenotype that produces cytokines and chemokines that maintain the inflammatory state, M2 macrophages are an anti-inflammatory phenotype that produces cytokines/chemokines anti-inflammatory and growth factors, and higher expression of M2 plays a role in the resolution of inflammation, antifibrotic activity and regeneration of tissues [10]. Inflammation in wound healing is also regulated by nuclear factors such as NFκB (proinflammatory) and Nrf2 (homeostatic and antioxidant response), and the regulation of these factors modulates inflammation in the wound, which promotes tissue regeneration [11]. However, more research is still needed in this regard.

If an active principle has the property of decreasing the inflammation time, without inhibiting this process, growth factors are produced that allow the healing stages to occur in less time (Table **1**).

Table 1. Extracts of plant species with wound regeneration effects.

Species	Extract and Plant Structure	Biological Model and Method	Compounds	Mechanism of Action	Refs.
Aloe vera Synonym *A. barbadensis*	Gel Leaves	Wistar rats (body weight 200-300 g) Excision wound model	Anthraquinones	In 21 days: re-epithelialized with a basal membrane ↑vascularized and well-organized collagen	[12]
A. vera Synonym *A. barbadensis*	Gel Leaves	Wistar rats (adult mature male, weighing 250 g) Excision wound model	Acemannan, maloyl glucans, aloine, emodin, anthrones, asarabinan, arabinorhamnogalactan, galactan, galactogalacturan, glucogalactomannan, galacto glucoarabinomannan, glucuronic acid, and lectins	modulated the inflammation, ↑rate and quality of fibroplasia, ↑remodeling stage, ↑ wound contraction, ↑epithelialization, ↑higher tissue alignment	[13]
Synonym *A. barbadensis*	Gel Leaves	Wistar male rats (250-300 g) Incision Wound Model	NR	↑regeneration epithelium ↑angiogenesis	[14]
Curcuma longa	Ethanolic Rhizome	Female rabbit *Oryctolagus cuniculus* (1500-2000 g) Incision Wound Model	Alkaloids, flavonoids, curcumin, essential oils, saponins, tannins, and terpenoids	↓decreases length of wound	[15]
Panax gingseng	Aqueous Roots	Male Sprague Dawley rats (200-250 g) Excision wound model	Ginsenoside	↑angiogenesis ↑epithelialization ↓ inflammatory cells ↑ number of fibroblasts ↑collagen	[16]
Artemisia montana	Essential oil Flowers	Human keratinocytes (HaCaT) Proliferation and migration	Major compounds in essential oil: β-caryophyllene (12.8%),	↑proliferation ↑synthesis of collagen, ↑wound closure.	[17]

(Table 1) cont.....

Species	Extract and Plant Structure	Biological Model and Method	Compounds	Mechanism of Action	Refs.
-	-	Sprague-Dawley rats (200-220 g) Excision wound model	germacrene D (9.9%), 1,8-cineole (7.9%), and camphor (6.2%)	-	-
Salvia officinalis	Hydroethanolic Leaves	White Wistar rats (200 g) Excision and incision wound models	Flavonoids and phenols	↑contraction and re-epithelialization in wound ↑new blood vessels and fibroblasts distribution	[18]
Garcinia brasiliensis	Leaves	Wistar rats (250-350 g) Excision wound model	Catechin, quercetin, tannins, berberine and garcinol	↑re-epithelialization ↓less edema better reorganization of the dermis	[19]
Calendula officinalis	Ethanolic and aqueous Flowers	BALB/c mice and primary human dermal fibroblasts (HDF) Excision wound model	Rutin and quercetin- 3-O-glucoside	HDF *in vitro*: ↑proliferation and migration of primary human dermal fibroblasts (HDF) ↑expression of connective tissue growth factor (CTGF) and α-smooth muscle actin (α-SMA) *In vivo:* ↑wound healing ↑ expression of CTGF and α-SMA	[20]
Portulaca oleracea	Ethanolic and aqueous fraction Leaves	Male C57BL/6J mice (9 weeks, 20-22 g) Excision wound model Human keratinocytes (HaCaT) Human endothelial cells (HUVEC) Cell migration assay	NR	↑proliferation of HUVEC ↑re-epithelialization pro-regenerative promotion of vascular and epidermal appendage ↑production of collagen	[21]

(Table 1) cont.....

Species	Extract and Plant Structure	Biological Model and Method	Compounds	Mechanism of Action	Refs.
Asterohyptis stellulata	Methanolic extract Aerial parts	Male CD1 et/et mice Incisional wound model	Quercetin derivatives and a glycoside rosmarinic acid	↑closure speed of the wound ↑increased tensile strength promoted mice skin regeneration, epidermal appendage	[22]

NR: Not reported; ↑: increase; ↓: decrease.

The active principles that present the mechanisms of action that have been mentioned produce a partial effect in tissue repair with the consequent formation of scars, while others may have regenerative properties through the proliferation of keratinocytes [17], of which epidermal appendages, as well as contraction and closure of the wound without the formation of an apparent scar. These effects are mainly because the extracts and/or active principles increase the expression of transformant growth factor-B (TGF-B) and vascular endothelial growth factor (VEGF), which have an important role in stimulating re-epithelialization, granulation tissue, angiogenesis, and the deposition of collagen fibers in incisional and excisional wounds [7], so such substances, whether extracts or pure compounds, can be considered to have a regenerative effect (Table 1). Additionally, the PI3K/AKT, ERK, and AKT/mTOR pathways participate in wound healing by promoting cell proliferation differentiation and migration [23, 24]. Its role in the wound regeneration process has been reported, but its role in skin regeneration is still being investigated.

The diversity of species that produce active principles with healing properties is enormous and they are part of the flora with medicinal properties, which is used throughout the world. Of special importance are the plants used in the countries of Asia, Africa, and America, where species such as: *Aloe vera* and the synonym *A. barbadensis* [7, 25], *Centella asiatica* [26], *Curcuma longa* [15], *Panax gingseng* [16], *Salvia officinalis* [18], *Asterohyptis stellulata* [22], among others (Table 1).

One of the most studied plants is *A. vera* (Synonym: *A. barbadensis*), which contains more than 200 active compounds such as anthraquinones -aloin, emodin, and chrysophanol-, anthrones, flavones, chromones, alkaloids, carbohydrates, amino acids, lipids, minerals, and vitamins, which may even act synergistically way. *A. vera* gel and extract promote skin repair in both *in vitro* and *in vivo* models [7, 21]. In addition, it has antioxidant, anti-inflammatory, antimicrobial and immunomodulatory properties [27]. These properties favor epithelialization, the organization of the ECM fibers, vascularization, and the remodeling of the skin and epidermal appendage, which is why it favors skin regeneration [7, 12, 14, 27, 28].

The compounds with regenerative capacity are diverse among the different plant species, such as phenolic compounds, alkaloids, glycosides, and glycoproteins (Table **2**). Of these, phenolic compounds [29], glycosides, and glycoproteins are especially relevant, which act by activating the growth factors involved in keratinocyte proliferation, angiogenesis, production and maturation of collagen, and wound closure [30 - 32].

Whether they are crude extracts, mixtures of compounds and/or pure active principles, the concentrations in which they exert the regenerative effect play a fundamental role, since in several of them, it has been observed that the effect is not always dependent on the concentration, in such a way that in lower concentrations of natural products greater healing is observed, which is a hormetic effect. This effect occurs in processes related to cell keratinocyte proliferation, viability, migration, and collagen deposition in murine and human fibroblasts [33].

Asiaticoside and madecassoside are pentacyclic triterpene glycosides isolated from *Centella asiatica*. Glycosides and their aglycones -asiatic acid and madecassic acid- are recognized for their beneficial effects on the skin. These compounds favor skin regeneration because they stimulate the synthesis of collagen, in addition to moisturizing the wound [34]. Madecassol®, contains the following active ingredients: asiaticoside, asiatic acid and madecassic. These compounds improve tensile strength, collagen synthesis, re-epithelialization, and promote angiogenesis [35 - 37]. The activity of these compounds has been evaluated in different models that include burns [38] as well as excisional wounds [39]. Curcumin (diferuloylmethane) is the main curcuminoid in *Curcuma longa*. This compound has antioxidant, antimicrobial, and anti-inflammatory properties. It favors the development of various phases of healing, especially the transition from the inflammatory to the proliferative phase, which causes wound contraction and subsequent remodeling, high collagen deposition, neovascularization, rapid re-epithelialization, and tissue formation [40]. The effect of various secondary metabolites on wound repair is described in Table **2**.

BIOTECHNOLOGY IN WOUND REGENERATION

The activity of natural products on the skin is limited, among other factors, by poor solubility in water, high hydrophobicity y photosensitivity. Various formulations are being developed to favor the bioavailability of substances by improving solubility, and permeability on the skin, their avoided degradation (chemical and/or enzymatic), promoting cell proliferation and their antibacterial and anti-inflammatory activities (Fig. **2**).

Table 2. Compounds with wound regeneration effects.

Compound	Plant Species	Biological Model	Mechanism of Action	Refs.
Acemannan	*Aloe vera* Synonym *A. barbadensis*	Male Sprague Dawley rats (8-week-old, 200-250 g) wound healing assay	↑ cell proliferation ↑KGF-1, VEGF, and type I collagen ↓ wound area ↑ epithelial coverage	[41]
Asiaticoside and madecassoside	*Centella asiatica*	Male Sprague-Dawley rats (250-300 g) Burn wound with hot plate (75 ℃) (3.5x4.6 cm)	↑ collagen synthesis ↓ oxidative stress ↑vasodilatation, cell proliferation and growth ↑VEGF and FGF	[31]
Madecassic acid, madecassoside, asiatic acid, and asiaticoside	*Centella asiatica*	Male ICR mice 12 weeks (20-25 g) Excision wound model	↑wound repair ↑fibroblast proliferation and collagen synthesis ↑contraction and epithelization	[39]
3α-hydroxymasticadienoic acid (3MA), masticadienoic acid (MA), anacardic acid (ANA)	*Amphipterygium adstringens*	Male Wistar rats (170-200 g) Excision wound model	3MA and ANA: ↑wound closure better collagen matrix architecture ANA and MA: ↓inflammatory infiltrate ↑mature epithelium and dermal papillae	[42]
3'4-O-dimethylcedrusin and proanthocyanidins	*Croton* spp	Female Wistar rats (250-300 g) Excision wound model	↑contraction of the wound ↑new collagen regeneration of the epithelial layer stimulation of fibroblast spread	[43]
Curcumin	*Curcuma longa*	-	↑granulation tissue ↑collagen deposition, remodeling and wound contraction	[44]
Rosmarinic acid	*Rosmarinus officinalis*	Sprague Dawley rats (180-220 g) Excision wound model	↑wound contraction ↑cell adhesion, epithelial migration, and high hydroxyproline content	[45]

(Table 2) cont.....

Compound	Plant Species	Biological Model	Mechanism of Action	Refs.
Epigallocatechin gallate	*Camellia sinensis*	Wistar rat (180-220 g and 3-5 weeks of age) Excision wound model	↑wound regeneration process ↑vascularization, modulation growth factors and inflammatory cytokines	[46]
Hesperidin	-	Adult Sprague Dawley rats (male, 180-220 g) diabetics Excision wound model	↓VEGF-c, Ang-1/Tie-2, TGF-β and Smad-2/3 ↑angiogenesis and vasculogenesis	[47]
Taxifolin	-	Male Wistar rats (250-300 g) Burn wound	regeneration and reparation of hair follicles and sebaceous glands	[48]
Naringin	-	Adult male Wistar rats (250-270 g) Excision wound model	Naringin 6% downregulates the expression of inflammatory factors (NF-κB, TNF-α and interleukins) and apoptotic factors (pol-γ and BAX); upregulates the expression of VEGF and TGF-β1, which are associated with the enhancement of collagen I synthesis and angiogenesis	[49]
Quercetin	-	Human skin fibroblasts (HSF), mouse skin fibroblasts (MSF) C57BL/6 mice	↑proliferation and migration of fibroblasts; restores the content of collagen fibers, ↓inflammatory factors, (TNF-α, IL-1β and IL-6) ↑ VEGF, FGF and SMA	[50]

NR: Not reported; ↑: increase; ↓: decrease.

Hydrogels

One technological strategy in regenerative medicine is the use of formulations/scaffolds for dressing and adjuvant drugs, mostly hydrogels, in the treatment of wound regeneration [51]. The hydrogel is a material of three-dimensional networks whose main characteristic is that it retains large amounts of water without dissolving, thus maintaining a humid environment that helps reduce the formation of scars. Additionally, high porosity allows gas exchange, and hydrogels can be loaded with bioactive molecules, in addition low adherence reduces pain for patients [52]. Therefore, the investigation of hydrogels loaded with extracts or secondary metabolites is essential for wound regeneration. Quercetin is a flavonoid that decreases fibrosis and scar formation, when loaded into liposomes and integrated in hydrogels, enhanced its bioavailability and maintaining its qualities [53]. Moreover, hydrogels enable the sustained release of metabolites. For example, polyvinyl alcohol hydrogel with inebrin of *Lagochilus inebrians* has a sustained release of the drug for up to 60 hours [54]. The combination of resveratrol and curcumin in a nanoemulgel enhanced controlled release, and it was applied in burns and showed histopathological features similar to those of normal skin [55]. Secondary metabolites also function as scaffolds for the formulation of biomaterials such as hydrogels.

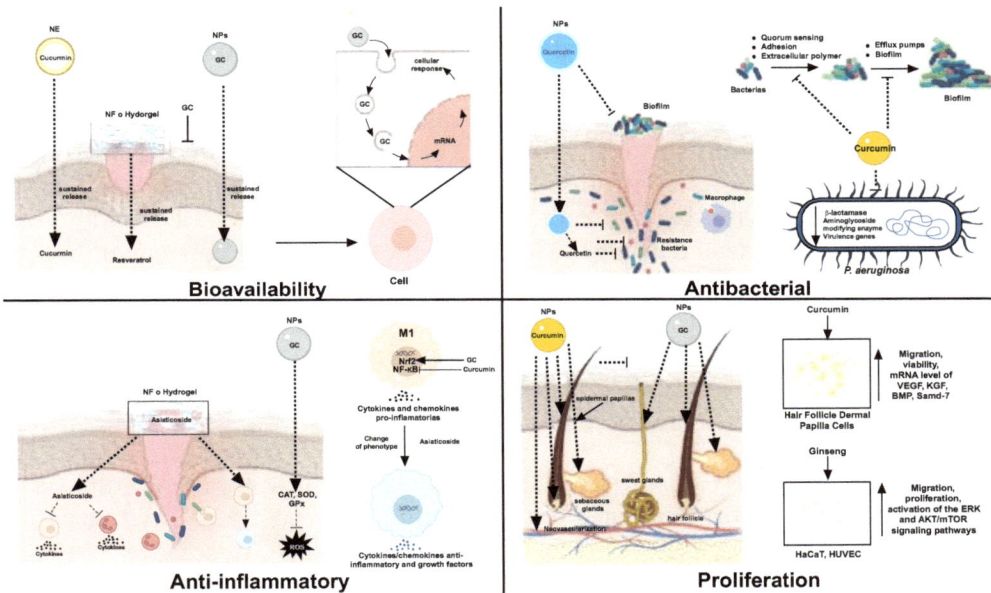

Fig. (2). Effect of biotechnological formulations on the penetration and bioavailability of different secondary metabolites and their effect on skin regeneration. CAT: Catalase, GC: Gallocathequin; GPx: Glutathione peroxidase, M1: type 1 macrophage, M2: type 2 macrophage, NE: Nanoemulsion, NPs: Nanoparticles, ROS: Reactive oxygen species, SOD: Superoxide dismutase. Created in BioRender.

Resveratrol in a gallic acid aqueous solution, promotes self-assembly of gallic acid to form a fibrous hydrogel that is also a therapeutic carrier and shows slow release, antibacterial effects, and regulation of inflammatory factors in bacteria-infected wounds [56]. Tannic acid, acrylamide, and soy protein constituting the hydrogel used to treat wounds exhibited good adhesion, antibacterial activity, and tissue regeneration [57].

Nanotechnology

Nanotechnology is a field in science involving the manipulation of matter in the nanoscale range (10-1000 nm), with different approaches such as medical area [58]. Their use as a carrier enhances bioavailability, increases biological activity and allows the use of a lower dose of the active substance [59]. Nanotechnology itself applied in wounds enhanced healing and regeneration. In the treatment of regeneration of the skin, the use of nanoparticles predominates, mainly applied together with hydrogels to a lesser extent and other nanomaterials designed to function as dressings.

Nanoparticles

Nanoparticles (NPs) in chronic wounds promote regeneration of skin, as shown by epidermal appendage formation -hair follicles, sweat and sebaceous glands-, formation of dermal papillae, adipocyte migration, and inhibition of scar [60, 61]. Moreover, NPs as administration vehicles keep the drug isolated from possible bioreactions and allow its controlled release after reaching the lesions [62], and NPs can enter the tissue by hydrophobic interactions and cells by endocytosis, acting as vehicle and drug delivery [63]. Skin acts as a natural barrier; in this case, it prevents most drugs from penetrating, mainly hydrophobic molecules with a large molecular weight [64]. Moreover, when the molecules pass through the stratum corneum, they are more retained in the dermis and hypodermis [65, 66]. Therefore, developing methods to avoid the stratum corneum and facilitate drug entry is essential in topical administration [67].

The modifications in NPs are essential since they will allow to penetrate tissues such as the stratum corneum or even determine the specificity of the cellular target. In AuNPs, their size, shape, and modifications of surface chemistry (charges, ligands, antibodies, polymers), determine whether penetration may occur through skin (hair follicles, intercellularly, or transcellularly), as well as release in response to pH, biomolecules, or light [68].

Medicinal plants and metabolite secondaries, as mentioned earlier, also have activities related to enhanced wound healing, promote the regeneration of skin and epidermal appendages. This activity is enhanced by their integration of NPs. *Cimifuga dahurica* in AgNPs alone and in hydrogel promotes healing skin wounds and formation of epidermal appendages [69]. *Calendula officinalis* is widely known for its healing activity; when it is integrated with AgNPs in chitosan hydrogels and applied in patients with very large ulcers (10 x 8 cm approximately) in the lower extremities that do not respond to treatment, decreased inflammation and pain occur, wounds heal completely and promote tissue regeneration [70].

The possible improvement in the activity of the extracts when integrated into NPs and/or hydrogels could be because of the increased bioavailability of the secondary metabolites. Some of these compounds have low bioavailability, skin absorption and sustained release problems. For example, flavonoids have a high molecular weight more when glycosylated, aromatic ring that confers hydrophobicity and steric hindrances, has little probability of transport across cell membranes, and the aglycones enter by passive diffusion but are pumped out by ABC transporters [71]. When flavonoids are integrated into NPs and/or hydrogels, their bioavailability improves.

Gallocatechin (GC) integrated into AgNPs increased wound healing, restauration of the dermis and epidermal appendages, enhanced the expression of antioxidant enzymes, and growth factors [11], increased Nrf2 and regulated the expression of NFκB; both of which are associated with changes in the phenotype of M1 to M2 macrophages [11]. Higher expression of M2 plays a role in the resolution of inflammation, antifibrotic activity and regeneration of tissues [10]. As mentioned, curcumin has various properties that favor regeneration wound -antibacterial, antioxidant and anti-inflammatory (NF-κB) activities- but has poor solubility and low bioavailability [72]. However, their bioavailability is enhanced when integrated with the NPS of Fe-SiO$_2$ or NPS of silk fibroin/alginate. They inhibit scar formation and promote follicle regeneration in bacterial-infected wound; and also increase cell migration and production of growth factors in hair follicles in the DP [73, 74]. In this case, the different material of NPs does not affect the activity since both improved tissue regeneration.

Hence, the use and application of NPs increase the liberation of drugs and prolong the release of metabolites. Alginate nanoparticles as capsaicin vehicle (embedded in nanofibers) prolonged the release of metabolites from 120 to 500 hours in the treatment of cancer [75]. Triterpene saponins such as asiaticoside are poorly soluble and lipophilic but are loaded in polymeric NPs and incorporated in hydrogel to improve bioavailability. They enable sustained release three times more that metabolites alone and enhance therapeutic effects in diabetic wounds [76]. Moreover, asiaticoside loading into polylactic-co-glycolic acid nanofibers promotes M2 polarization, and downregulates inflammatory cytokines, which could promote tissue regeneration [77]. Secondary metabolites yields are influenced by biotic and abiotic factors, and industrial manufacturing is complicated due to its chemical complexity. Therefore, it is especially important make the most of the resources.

The widespread use of antibiotics has increased antimicrobial resistance; for example, the activity of β-lactamase enzymes and formation of biofilms. Nanomaterials exhibit antibacterial properties, which accelerate wound regeneration [78]. AgNPs have notorious antibacterial activities *per se* against gram-negative and gram-positive.

Secondary metabolites act on bacteria then: 1) compromise the cell wall/membrane; 2) inhibit the production of biomolecules; and c) cause death by ROS [79]. Moreover, wound dressings loaded with AgNPs have antimicrobial activity that downregulates β-lactamase and other resistance genes [80]; as previously mentioned, by integrating bioactive compounds such as GC, they act synergistically and favor skin regeneration [11]. Quercetin-borate NPs in poly vinyl alcohol exhibited excellent bacteriostasis and functional restoration of skin [60]. Biofilms inhibit the healing of chronic wounds; treatment to remove them is complicated by the presence of an extracellular polymeric barrier (EPB) that is permeable to drugs [81]. Nanomaterials, owing to their size, can penetrate and deliver antibacterial agents in tissues and biofilms [82]. Sonodynamic therapy with emodin NPs inhibited the growth of multi-bacterial biofilms, and curcumin NPs in biofilms of multidrug-resistant *Pseudomonas aeruginosa*, both burn wound bacterial isolates, had antibiofilm activity by decreasing expression of virulence factors [83, 84]. Therefore, the formulation with nanoparticles and natural products could be a good alternative for the treatment of wounds with bacterial resistant and biofilm, promoting of tissue regeneration.

Metabolites are not only being used as bioactive compounds but also for the construction of NPs. Glycosylated flavonoids such as rutin are used as scaffolds of NPs (rutin NPs) incorporated into cryogels to increased IL-10, have antioxidant properties, and demonstrate their antiscarring effect [85]. Another example is the fabrication of NPs with Ginseng extract which increased cell migration and proliferation (ERK, AKT/mTOR), angiogenesis, dermis similar to normal and with hair growth [86].

Nanofibrous

Another application of nanotechnology for regenerative medicine is the design of biomaterials that provide essential structural "scaffolding" to create new structures that mimic natural tissues that are biocompatible [87]. Nanofibrous scaffolds (NFS) or matrix is a biomaterial that forms a 3D framework composed of synthetic or natural polymer to carry and deliver drugs facilitating tissue regeneration in wounds [60]. Investigation of this material with extract or metabolites is still limited because most of the research that have been done only reach the *in vitro* phase. *Aloe vera* loaded in NFS zein/polycaprolactone/collagen had antibacterial activity against *S. aureus* and *E. coli* [88]. Essential oils have excellent biological properties, but low solubility and stability under external factors limit their use [59]. Integration in NFS enhanced its stability. Polyvinyl alcohol/gelatin NFS with *Thymus daenesis* essential oil and *Glycyrrhiza glabra* extract promote the migration of fibroblasts [89]. NFS with quercetin in graphene oxide NPs and incorporation in polycaprolactone solution promoted the liberation of quercetin by approximately 70% after 15 days and had an antibacterial effect [90].

Nanoemulsion

Nanoemulsion is a mix of two immiscible liquids in the presence of surfactants; one of the liquids disperses in another liquid and forms small spherical droplets [72]. In the case of skin, monoterpenes and sesquiterpenes penetrate fastest and deeper, this is time-dependent and these compounds enhance the chemical absorption of other triterpenes that are more complex and do not penetrate the skin easily [91, 92]. However, when integrated into a nanoemulsion, triterpenes improve bioavailability. For example, a nanoemulsion with triterpenes extract of *Poria cocos* loaded in hydrogels for topical application stimulates skin regeneration in diabetic wounds [93]. Nanoemulsion of curcumin induces skin regeneration [72].

CONCLUSION

When the integrity of the skin is lost and a wound occurs, a repair process begins that can restore the functionality of this organ through two mechanisms: healing with the formation of a fibrotic scar or skin regeneration where the skin annexes are recovered, such as hair follicles and glands. The occurrence of either mechanism depends on a complex regulation process that is still not fully understood, this process involves molecular mechanisms that activate and regulate re-epithelialization, formation of epidermal appendages, and neovascularization.

Plant extracts and their active components tip the balance toward regeneration; however, bioavailability in many cases is limited. The development of hydrogels and various nanotechnological strategies -nanoparticles, nanofibers and nanoemulsions- promote the penetration, bioavailability, and improvement of the therapeutic effect of plants.

In many cases, the low doses of the active compounds reduce the possible risk of presenting side effects. Therefore, the use of technology integrating natural products is an important point within research to find effective treatments for chronic wounds that can lead to loss of tissue function, amputation of limbs and even death. However, it is essential to continue researching natural products to understand how they promote the regeneration of skin wounds.

ACKNOWLEDGEMENT

DGAPA, PAPIIT-UNAM-IN212623, Biological Sciences Postgraduate Studies Unit-UNAM, and Doctorate's Scholarship-CONACYT (CVU 775307).

REFERENCES

[1]   Woodley DT. Distinct fibroblasts in the papillary and reticular dermis: Implications for wound healing. Dermatol Clin 2017; 35(1): 95-100.
        [http://dx.doi.org/10.1016/j.det.2016.07.004] [PMID: 27890241]

[2]   Rippa AL, Kalabusheva EP, Vorotelyak EA. Regeneration of dermis: Scarring and cells involved. Cells 2019; 8(6): 607.
        [http://dx.doi.org/10.3390/cells8060607] [PMID: 31216669]

[3]   Reinke JM, Sorg H. Wound repair and regeneration. Eur Surg Res 2012; 49(1): 35-43.
        [http://dx.doi.org/10.1159/000339613] [PMID: 22797712]

[4]   Phan QM, Sinha S, Biernaskie J, Driskell RR. Single cell transcriptomic analysis of small and large wounds reveals the distinct spatial organization of regenerative fibroblasts. Exp Dermatol 2021; 30(1): 92-101.
        [http://dx.doi.org/10.1111/exd.14244] [PMID: 33237598]

[5]   Jiang D, Rinkevich Y. Scars or regeneration? Dermal fibroblasts as drivers of diverse skin wound responses. Int J Mol Sci 2020; 21(2): 617.
        [http://dx.doi.org/10.3390/ijms21020617] [PMID: 31963533]

[6]   Lim CH, Sun Q, Ratti K, *et al.* Hedgehog stimulates hair follicle neogenesis by creating inductive dermis during murine skin wound healing. Nat Commun 2018; 9(1): 4903.
        [http://dx.doi.org/10.1038/s41467-018-07142-9] [PMID: 30464171]

[7]   Yazarlu O, Iranshahi M, Kashani HRK, *et al.* Perspective on the application of medicinal plants and natural products in wound healing: A mechanistic review. Pharmacol Res 2021; 174: 105841.
        [http://dx.doi.org/10.1016/j.phrs.2021.105841] [PMID: 34419563]

[8]   Bao L, Cai X, Zhang M, *et al.* Bovine collagen oligopeptides accelerate wound healing by promoting fibroblast migration *via* PI3K/Akt/mTOR signaling pathway. J Funct Foods 2022; 90: 104981.
        [http://dx.doi.org/10.1016/j.jff.2022.104981]

[9]   Liu C, Zeng H, Chen Z, *et al.* Sprayable methacrylic anhydride-modified gelatin hydrogel combined with bionic neutrophils nanoparticles for scar-free wound healing of diabetes mellitus. Int J Biol Macromol 2022; 202: 418-30.
        [http://dx.doi.org/10.1016/j.ijbiomac.2022.01.083] [PMID: 35051497]

[10]   Monavarian M, Kader S, Moeinzadeh S, Jabbari E. Regenerative Scar Free Skin Wound Healing. Tissue Eng Part B Rev 2019; 25(4): 294-311.
        [http://dx.doi.org/10.1089/ten.teb.2018.0350] [PMID: 30938269]

[11]   Vendidandala NR, Yin TP, Nelli G, Pasupuleti VR, Nyamathulla S, Mokhtar SI. Gallocatechin silver nanoparticle impregnated cotton gauze patches enhance wound healing in diabetic rats by suppressing oxidative stress and inflammation *via* modulating the Nrf2/HO-1 and TLR4/NF-κB pathways. Life Sci 2021; 286: 120019.
        [http://dx.doi.org/10.1016/j.lfs.2021.120019] [PMID: 34624322]

[12]   Brandão ML, Reis PRM, Araújo LA, Araújo ACV, Santos MHAS, Miguel MP. Evaluation of wound healing treated with latex derived from rubber trees and *Aloe Vera* extract in rats. Acta Cir Bras 2016; 31(9): 570-7.
        [http://dx.doi.org/10.1590/S0102-865020160090000001] [PMID: 27737341]

[13]   Oryan A, Mohammadalipour A, Moshiri A, Tabandeh MR. Topical application of *Aloe vera* accelerated wound healing, modeling, and remodeling: An experimental study. Ann Plast Surg 2016;

77(1): 37-46.
[http://dx.doi.org/10.1097/SAP.0000000000000239] [PMID: 25003428]

[14]   Tarameshloo M, Norouzian M, Zarein-Dolab S, Dadpay M, Mohsenifar J, Gazor R. *Aloe vera* gel and thyroid hormone cream may improve wound healing in Wistar rats. Anat Cell Biol 2012; 45(3): 170-7.
[http://dx.doi.org/10.5115/acb.2012.45.3.170] [PMID: 23094205]

[15]   Adeliana U, Ahmad AN, Arifuddin S, Yulianty R. Prihantono. Effectiveness of turmeric (*Curcuma longa* Linn) gel extract (GE) on wound healing: Pre-clinical test. Gac Sanit 2021; S2: S196-8.

[16]   Park KS, Park DH. The effect of Korean Red Ginseng on full-thickness skin wound healing in rats. J Ginseng Res 2019; 43(2): 226-35.
[http://dx.doi.org/10.1016/j.jgr.2017.12.006] [PMID: 30976160]

[17]   Yoon MS, Won KJ, Kim DY, *et al.* Skin regeneration effect and chemical composition of essential oil from *Artemisia montana.* Nat Prod Commun 2014; 9(11): 1934578X1400901.
[http://dx.doi.org/10.1177/1934578X1400901123] [PMID: 25532296]

[18]   Karimzadeh S, Farahpour MR. Topical application of *Salvia officinalis* hydroethanolic leaf extract improves wound healing process. Indian J Exp Biol 2017; 55(2): 98-106.
[PMID: 30183236]

[19]   Souza HR, Zucoloto AR, Francisco ITP, *et al.* Evaluation of the healing properties of *Garcinia brasiliensis* extracts in a cutaneous wound model. J Ethnopharmacol 2022; 295: 115334.
[http://dx.doi.org/10.1016/j.jep.2022.115334] [PMID: 35597412]

[20]   Dinda M, Mazumdar S, Das S, *et al.* The water fraction of *Calendula officinalis* hydroethanol extract stimulates *in vitro* and *in vivo* proliferation of dermal fibroblasts in wound healing. Phytother Res 2016; 30(10): 1696-707.
[http://dx.doi.org/10.1002/ptr.5678] [PMID: 27426257]

[21]   Guo J, Peng J, Han J, *et al.* Extracts of *Portulaca oleracea* promote wound healing by enhancing angiology regeneration and inhibiting iron accumulation in mice. Chin Herb Med 2022; 14(2): 263-72.
[http://dx.doi.org/10.1016/j.chmed.2021.09.014] [PMID: 36117668]

[22]   Álvarez-Santos N, Estrella-Parra EA, Benítez-Flores JC, *et al. Asterohyptis stellulata:* Phytochemistry and wound healing activity. Food Biosci 2022; 102150.

[23]   Wei X, Luo L, Chen J. Roles of mTOR signaling in tissue regeneration. Cells 2019; 8(9): 1075.
[http://dx.doi.org/10.3390/cells8091075] [PMID: 31547370]

[24]   Teng Y, Fan Y, Ma J, *et al.* The PI3K/Akt Pathway: Emerging roles in skin homeostasis and a group of non-malignant skin disorders. Cells 2021; 10(5): 1219.
[http://dx.doi.org/10.3390/cells10051219] [PMID: 34067630]

[25]   Moriyama M, Moriyama H, Uda J, *et al.* Beneficial effects of the genus *Aloe* on wound healing, cell proliferation, and differentiation of epidermal keratinocytes. PLoS One 2016; 11(10): e0164799.
[http://dx.doi.org/10.1371/journal.pone.0164799] [PMID: 27736988]

[26]   Arribas-López E, Zand N, Ojo O, Snowden MJ, Kochhar T. A systematic review of the effect of *Centella asiatica* on wound healing. Int J Environ Res Public Health 2022; 19(6): 3266.
[http://dx.doi.org/10.3390/ijerph19063266] [PMID: 35328954]

[27]   Liang J, Cui L, Li J, Guan S, Zhang K, Li J. *Aloe vera*: A medicinal plant used in skin wound healing. Tissue Eng Part B Rev 2021; 27(5): 455-74.
[http://dx.doi.org/10.1089/ten.teb.2020.0236] [PMID: 33066720]

[28]   Movaffagh J, Khatib M, Fazly Bazzaz BS, *et al.* Evaluation of wound-healing efficiency of a functional Chitosan/Aloe vera hydrogel on the improvement of re-epithelialization in full thickness wound model of rat. J Tissue Viability 2022; 31(4): 649-56.

[29]   Melguizo-Rodríguez L, de Luna-Bertos E, Ramos-Torrecillas J, Illescas-Montesa R, Costela-Ruiz VJ, García-Martínez O. Potential effects of phenolic compounds that can be found in olive oil on wound

healing. Foods 2021; 10(7): 1642.
[http://dx.doi.org/10.3390/foods10071642] [PMID: 34359512]

[30]   Choi SW, Son BW, Son YS, Park YI, Lee SK, Chung MH. The wound-healing effect of a
        glycoprotein fraction isolated from *Aloe vera*. Br J Dermatol 2001; 145(4): 535-45.
        [http://dx.doi.org/10.1046/j.1365-2133.2001.04410.x] [PMID: 11703278]

[31]   Zhang Y, Wang T, Zhang D, *et al.* Chitosan based macromolecular hydrogel loaded total glycosides of
        paeony enhances diabetic wound healing by regulating oxidative stress microenvironment. Int J Biol
        Macromol 2023; 250(1): 126010.

[32]   Kartini K, Wati N, Gustav R, *et al.* Wound healing effects of *Plantago major* extract and its chemical
        compounds in hyperglycemic rats. Food Biosci 2021; 41: 100937.
        [http://dx.doi.org/10.1016/j.fbio.2021.100937]

[33]   Calabrese EJ, Dhawan G, Kapoor R, Agathokleous E, Calabrese V. Hormesis: Wound healing and
        fibroblasts. Pharmacol Res 2022; 184: 106449.
        [http://dx.doi.org/10.1016/j.phrs.2022.106449] [PMID: 36113746]

[34]   Bandopadhyay S, Mandal S, Ghorai M, *et al.* Therapeutic properties and pharmacological activities of
        asiaticoside and madecassoside: A review. J Cell Mol Med 2023; 27(5): 593-608.
        [http://dx.doi.org/10.1111/jcmm.17635] [PMID: 36756687]

[35]   Maquart FX, Chastang F, Simeon A, Birembaut P, Gillery P, Wegrowski Y. Triterpenes from *Centella
        asiatica* stimulate extracellular matrix accumulation in rat experimental wounds. Eur J Dermatol 1999;
        9(4): 289-96.
        [PMID: 10356407]

[36]   Bylka W, Znajdek-Awiżeń P, Studzińska-Sroka E, Brzezińska M. *Centella asiatica* in cosmetology.
        Postepy Dermatol Alergol 2013; 1(1): 46-9.
        [http://dx.doi.org/10.5114/pdia.2013.33378] [PMID: 24278045]

[37]   Maramaldi G, Togni S, Franceschi F, Lati E. Anti-inflammaging and antiglycation activity of a novel
        botanical ingredient from African biodiversity (Centevita™). Clin Cosmet Investig Dermatol 2013; 7:
        1-9.
        [http://dx.doi.org/10.2147/CCID.S49924] [PMID: 24376360]

[38]   Hou Q, Li M, Lu YH, Liu DH, Li CC. Burn wound healing properties of asiaticoside and
        madecassoside. Exp Ther Med 2016; 12(3): 1269-74.
        [http://dx.doi.org/10.3892/etm.2016.3459] [PMID: 27588048]

[39]   Tanga BM, Bang S, Fang X, *et al.* RETRACTED: Centella asiatica extract in carboxymethyl cellulose
        at its optimal concentration improved wound healing in mice models. Heliyon 2022; 8(12): e12031.
        [http://dx.doi.org/10.1016/j.heliyon.2022.e12031] [PMID: 36531634]

[40]   Akbik D, Ghadiri M, Chrzanowski W, Rohanizadeh R. Curcumin as a wound healing agent. Life Sci
        2014; 116(1): 1-7.
        [http://dx.doi.org/10.1016/j.lfs.2014.08.016] [PMID: 25200875]

[41]   Jettanacheawchankit S, Sasithanasate S, Sangvanich P, Banlunara W, Thunyakitpisal P. Acemannan
        stimulates gingival fibroblast proliferation; expressions of keratinocyte growth factor-1, vascular
        endothelial growth factor, and type I collagen; and wound healing. J Pharmacol Sci 2009; 109(4): 525-
        31.
        [http://dx.doi.org/10.1254/jphs.08204FP] [PMID: 19372635]

[42]   Pérez-Contreras CV, Alvarado-Flores J, Orona-Ortiz A, *et al.* Wound healing activity of the
        hydroalcoholic extract and the main metabolites of *Amphipterygium adstringens* (cuachalalate) in a rat
        excision model. J Ethnopharmacol 2022; 293: 115313.
        [http://dx.doi.org/10.1016/j.jep.2022.115313] [PMID: 35461988]

[43]   Pieters L, De Bruyne T, Van Poel B, *et al. In vivo* wound healing activity of Dragon's Blood (*Croton
        spp.*), a traditional South American drug, and its constituents. Phytomedicine 1995; 2(1): 17-22.

[http://dx.doi.org/10.1016/S0944-7113(11)80043-7] [PMID: 23196095]

[44]   Tejada S, Manayi A, Daglia M, *et al.* Wound healing effects of curcumin: a short review. Curr Pharm Biotechnol 2016; 17(11): 1002-7.

[45]   Chhabra P, Chauhan G, Kumar A. Augmented healing of full thickness chronic excision wound by rosmarinic acid loaded chitosan encapsulated graphene nanopockets. Drug Dev Ind Pharm 2020; 46(6): 878-88.
        [http://dx.doi.org/10.1080/03639045.2020.1762200] [PMID: 32338544]

[46]   Kar AK, Singh A, Dhiman N, *et al.* Polymer-assisted *in situ* synthesis of silver nanoparticles with epigallocatechin gallate (EGCG) impregnated wound patch potentiate controlled inflammatory responses for brisk wound healing. Int J Nanomedicine 2019; 14: 9837-54.
        [http://dx.doi.org/10.2147/IJN.S228462] [PMID: 31849472]

[47]   Li W, Kandhare AD, Mukherjee AA, Bodhankar SL. Hesperidin, a plant flavonoid accelerated the cutaneous wound healing in streptozotocin-induced diabetic rats: Role of TGF-ß/Smads and Ang-1/Tie-2 signaling pathways. EXCLI J 2018; 17: 399-419.
        [PMID: 29805347]

[48]   Shubina VS, Shatalin YV. Skin regeneration after chemical burn under the effect of taxifolin-based preparations. Bull Exp Biol Med 2012; 154(1): 152-7.
        [http://dx.doi.org/10.1007/s10517-012-1897-z] [PMID: 23330113]

[49]   Salehi M, Vaez A, Naseri-Nosar M, *et al.* Naringin-loaded poly(ε-caprolactone)/gelatin electrospun mat as a potential wound dressing: *In vitro* and *in vivo* evaluation. Fibers Polym 2018; 19(1): 125-34.
        [http://dx.doi.org/10.1007/s12221-018-7528-6]

[50]   Mi Y, Zhong L, Lu S, *et al.* Quercetin promotes cutaneous wound healing in mice through Wnt/β-catenin signaling pathway. J Ethnopharmacol 2022; 290: 115066.
        [http://dx.doi.org/10.1016/j.jep.2022.115066] [PMID: 35122975]

[51]   Berce C, Muresan MS, Soritau O, *et al.* Cutaneous wound healing using polymeric surgical dressings based on chitosan, sodium hyaluronate and resveratrol. A preclinical experimental study. Colloids Surf B Biointerfaces 2018; 163: 155-66.
        [http://dx.doi.org/10.1016/j.colsurfb.2017.12.041] [PMID: 29291501]

[52]   Zhang X, Qin M, Xu M, *et al.* The fabrication of antibacterial hydrogels for wound healing. Eur Polym J 2021; 146: 110268.
        [http://dx.doi.org/10.1016/j.eurpolymj.2021.110268]

[53]   Jangde R, Srivastava S, Singh MR, Singh D. *In vitro* and *in vivo* characterization of quercetin loaded multiphase hydrogel for wound healing application. Int J Biol Macromol 2018; 115: 1211-7.
        [http://dx.doi.org/10.1016/j.ijbiomac.2018.05.010] [PMID: 29730004]

[54]   Djumaev A, Tashmukhamedova S. Physical and chemical properties of PVA-CMC based hydrogel carrier loaded with herbal hemostatic agent for application as wound dressings. Natl J Physiol Pharm Pharmacol 2020; 10(10): 905-9.

[55]   Alyoussef A, El-Gogary RI, Ahmed RF, Ahmed Farid OAH, Bakeer RM, Nasr M. The beneficial activity of curcumin and resveratrol loaded in nanoemulgel for healing of burn-induced wounds. J Drug Deliv Sci Technol 2021; 62: 102360.
        [http://dx.doi.org/10.1016/j.jddst.2021.102360]

[56]   Wang XC, Huang HB, Gong W, *et al.* Resveratrol triggered the quick self-assembly of gallic acid into therapeutic hydrogels for healing of bacterially infected wounds. Biomacromolecules 2022; 23(4): 1680-92.
        [http://dx.doi.org/10.1021/acs.biomac.1c01616] [PMID: 35258295]

[57]   Huang X, Ma C, Xu Y, *et al.* A tannin-functionalized soy protein-based adhesive hydrogel as a wound dressing. Ind Crops Prod 2022; 182: 114945.
        [http://dx.doi.org/10.1016/j.indcrop.2022.114945]

[58] Paiva-Santos AC, Herdade AM, Guerra C, *et al.* Plant-mediated green synthesis of metal-based nanoparticles for dermopharmaceutical and cosmetic applications. Int J Pharm 2021; 597: 120311.
[http://dx.doi.org/10.1016/j.ijpharm.2021.120311] [PMID: 33539998]

[59] Miastkowska M, Sikora E, Kulawik-Pióro A, *et al.* Bioactive *Lavandula angustifolia* essential oil-loaded nanoemulsion dressing for burn wound healing. *In vitro* and *in vivo* studies. Biomat Adv 2023; 148: 213362.
[http://dx.doi.org/10.1016/j.bioadv.2023.213362] [PMID: 36921462]

[60] Li X, Yang X, Wang Z, *et al.* Antibacterial, antioxidant and biocompatible nanosized quercetin-PVA xerogel films for wound dressing. Colloids Surf B Biointerfaces 2022; 209(Pt 2): 112175.
[http://dx.doi.org/10.1016/j.colsurfb.2021.112175] [PMID: 34740095]

[61] Chappidi S, Buddolla V, Ankireddy SR, Lakshmi BA, Kim YJ. Recent trends in diabetic wound healing with nanofibrous scaffolds. Eur J Pharmacol 2023; 945: 175617.
[http://dx.doi.org/10.1016/j.ejphar.2023.175617] [PMID: 36841285]

[62] McLean K, Zhan W. Mathematical modelling of nanoparticle-mediated topical drug delivery to skin tissue. Int J Pharm 2022; 611: 121322.
[http://dx.doi.org/10.1016/j.ijpharm.2021.121322] [PMID: 34848364]

[63] Shamiya Y, Ravi SP, Coyle A, Chakrabarti S, Paul A. Engineering nanoparticle therapeutics for impaired wound healing in diabetes. Drug Discov Today 2022; 27(4): 1156-66.
[http://dx.doi.org/10.1016/j.drudis.2021.11.024] [PMID: 34839040]

[64] Chen Y, Feng X, Meng S. Site-specific drug delivery in the skin for the localized treatment of skin diseases. Expert Opin Drug Deliv 2019; 16(8): 847-67.
[http://dx.doi.org/10.1080/17425247.2019.1645119] [PMID: 31311345]

[65] Sobanko JF, Miller CJ, Alster TS. Topical anesthetics for dermatologic procedures: A review. Dermatol Surg 2012; 38(5): 709-21.
[http://dx.doi.org/10.1111/j.1524-4725.2011.02271.x] [PMID: 22243434]

[66] Arora D, Nanda S. Quality by design driven development of resveratrol loaded ethosomal hydrogel for improved dermatological benefits *via* enhanced skin permeation and retention. Int J Pharm 2019; 567: 118448.
[http://dx.doi.org/10.1016/j.ijpharm.2019.118448] [PMID: 31226472]

[67] Elkomy MH, Ali AA, Eid HM. Chitosan on the surface of nanoparticles for enhanced drug delivery: A comprehensive review. J Control Release 2022; 351: 923-40.
[http://dx.doi.org/10.1016/j.jconrel.2022.10.005] [PMID: 36216174]

[68] Chen Y, Feng X. Gold nanoparticles for skin drug delivery. Int J Pharm 2022; 625: 122122.
[http://dx.doi.org/10.1016/j.ijpharm.2022.122122] [PMID: 35987319]

[69] Li Y, Sang Y, Yu W, Zhang F, Wang X. Antibacterial actions of Ag nanoparticles synthesized from Cimicifuga dahurica (Turcz.) Maxim. and their application in constructing a hydrogel spray for healing skin wounds. Food Chem 2023; 418: 135981.
[http://dx.doi.org/10.1016/j.foodchem.2023.135981] [PMID: 36996658]

[70] Rodríguez-Acosta H, Tapia- Rivera JM, Guerrero-Guzmán A, *et al.* Chronic wound healing by controlled release of chitosan hydrogels loaded with silver nanoparticles and *Calendula* extract. J Tissue Viability 2022; 31(1): 173-9.
[http://dx.doi.org/10.1016/j.jtv.2021.10.004] [PMID: 34774393]

[71] Kaushal N, Singh M, Singh Sangwan R. Flavonoids: Food associations, therapeutic mechanisms, metabolism and nanoformulations. Food Res Int 2022; 157: 111442.
[http://dx.doi.org/10.1016/j.foodres.2022.111442] [PMID: 35761682]

[72] Kumari M, Nanda DK. Potential of Curcumin nanoemulsion as antimicrobial and wound healing agent in burn wound infection. Burns 2023; 49(5): 1003-16.
[http://dx.doi.org/10.1016/j.burns.2022.10.008] [PMID: 36402615]

[73]   Babaluei M, Mottaghitalab F, Seifalian A, Farokhi M. Injectable multifunctional hydrogel based on carboxymethylcellulose/polyacrylamide/polydopamine containing vitamin C and curcumin promoted full-thickness burn regeneration. Int J Biol Macromol 2023; 236: 124005.
[http://dx.doi.org/10.1016/j.ijbiomac.2023.124005] [PMID: 36907296]

[74]   Jing Y, Ruan L, Jiang G, *et al.* Regenerated silk fibroin and alginate composite hydrogel dressings loaded with curcumin nanoparticles for bacterial-infected wound closure. Biomat Adv 2023; 149: 213405.
[http://dx.doi.org/10.1016/j.bioadv.2023.213405] [PMID: 37004308]

[75]   Ahmady AR, Solouk A, Saber-Samandari S, Akbari S, Ghanbari H, Brycki BE. Capsaicin-loaded alginate nanoparticles embedded polycaprolactone-chitosan nanofibers as a controlled drug delivery nanoplatform for anticancer activity. J Colloid Interface Sci 2023; 638: 616-28.
[http://dx.doi.org/10.1016/j.jcis.2023.01.139] [PMID: 36774875]

[76]   Narisepalli S, Salunkhe SA, Chitkara D, Mittal A. Asiaticoside polymeric nanoparticles for effective diabetic wound healing through increased collagen biosynthesis: *In-vitro* and *in-vivo* evaluation. Int J Pharm 2023; 631: 122508.
[http://dx.doi.org/10.1016/j.ijpharm.2022.122508] [PMID: 36539166]

[77]   Huang J, Zhou X, Shen Y, *et al.* Asiaticoside loading into polylactic co glycolic acid electrospun nanofibers attenuates host inflammatory response and promotes M2 macrophage polarization. J Biomed Mater Res A 2020; 108(1): 69-80.
[http://dx.doi.org/10.1002/jbm.a.36793] [PMID: 31496042]

[78]   Ehtesabi H, Fayaz M, Hosseini-Doabi F, Rezaei P. The application of green synthesis nanoparticles in wound healing: A review. Mat Tod Sustain 2023; 21: 100272.
[http://dx.doi.org/10.1016/j.mtsust.2022.100272]

[79]   Asghar S, Khan IU, Salman S, Khalid SH, Ashfaq R, Vandamme TF. Plant-derived nanotherapeutic systems to counter the overgrowing threat of resistant microbes and biofilms. Adv Drug Deliv Rev 2021; 179: 114019.
[http://dx.doi.org/10.1016/j.addr.2021.114019] [PMID: 34699940]

[80]   El-Aassar MR, Ibrahim OM, Fouda MMG, *et al.* Wound dressing of chitosan-based-crosslinked gelatin/polyvinyl pyrrolidone embedded silver nanoparticles, for targeting multidrug resistance microbes. Carbohydr Polym 2021; 255: 117484.
[http://dx.doi.org/10.1016/j.carbpol.2020.117484] [PMID: 33436244]

[81]   Yu X, Zhao J, Ma X, Fan D. A multi-enzyme cascade microneedle reaction system for hierarchically MRSA biofilm elimination and diabetic wound healing. Chem Eng J 2023; 465: 142933.
[http://dx.doi.org/10.1016/j.cej.2023.142933]

[82]   Razdan K, Garcia-Lara J, Sinha VR, Singh KK. Pharmaceutical strategies for the treatment of bacterial biofilms in chronic wounds. Drug Discov Today 2022; 27(8): 2137-50.
[http://dx.doi.org/10.1016/j.drudis.2022.04.020] [PMID: 35489675]

[83]   Shariati A, Asadian E, Fallah F, *et al.* Evaluation of Nano-curcumin effects on expression levels of virulence genes and biofilm production of multidrug-resistant *Pseudomonas aeruginosa* isolated from burn wound infection in Tehran, Iran. Infect Drug Resist 2019; 12: 2223-35.
[http://dx.doi.org/10.2147/IDR.S213200] [PMID: 31440064]

[84]   Pourhajibagher M, Rahimi-esboei B, Ahmadi H, Bahador A. The anti-biofilm capability of nano-emodin-mediated sonodynamic therapy on multi-species biofilms produced by burn wound bacterial strains. Photodiagn Photodyn Ther 2021; 34: 102288.
[http://dx.doi.org/10.1016/j.pdpdt.2021.102288] [PMID: 33836275]

[85]   Saha R, Patkar S, Pillai MM, Tayalia P. Bilayered skin substitute incorporating rutin nanoparticles for antioxidant, anti-inflammatory, and anti-fibrotic effect. Biomat Adv 2023; 150: 213432.
[http://dx.doi.org/10.1016/j.bioadv.2023.213432] [PMID: 37119696]

[86]   Yang S, Lu S, Ren L, *et al.* Ginseng-derived nanoparticles induce skin cell proliferation and promote wound healing. J Ginseng Res 2023; 47(1): 133-43.
[http://dx.doi.org/10.1016/j.jgr.2022.07.005] [PMID: 36644388]

[87]   Haleem A, Javaid M, Singh RP, Rab S, Suman R. Applications of nanotechnology in medical field: A brief review. Global Health Journal 2023; 7(2): 70-7.
[http://dx.doi.org/10.1016/j.glohj.2023.02.008]

[88]   Ghorbani M, Nezhad-Mokhtari P, Ramazani S. *Aloe vera*-loaded nanofibrous scaffold based on Zein/Polycaprolactone/Collagen for wound healing. Int J Biol Macromol 2020; 153: 921-30.
[http://dx.doi.org/10.1016/j.ijbiomac.2020.03.036] [PMID: 32151718]

[89]   Reisi-Vanani V, Hosseini S, Soleiman-Dehkordi E, *et al.* Engineering of a core-shell polyvinyl alcohol/gelatin fibrous scaffold for dual delivery of *Thymus daenensis* essential oil and *Glycyrrhiza glabra* L. extract as an antibacterial and functional wound dressing. J Drug Deliv Sci Technol 2023; 81: 104282.
[http://dx.doi.org/10.1016/j.jddst.2023.104282]

[90]   Faraji S, Nowroozi N, Nouralishahi A, Shabani Shayeh J. Electrospun poly-caprolactone/graphene oxide/quercetin nanofibrous scaffold for wound dressing: Evaluation of biological and structural properties. Life Sci 2020; 257: 118062.
[http://dx.doi.org/10.1016/j.lfs.2020.118062] [PMID: 32652138]

[91]   Cal K, Kupiec K, Sznitowska M. Effect of physicochemical properties of cyclic terpenes on their *ex vivo* skin absorption and elimination kinetics. J Dermatol Sci 2006; 41(2): 137-42.
[http://dx.doi.org/10.1016/j.jdermsci.2005.09.003] [PMID: 16260121]

[92]   Tang Q, Xu F, Wei X, *et al.* Investigation of β-caryophyllene as terpene penetration enhancer: Role of *Stratum corneum* retention. Eur J Pharm Sci 2023; 183: 106401.
[http://dx.doi.org/10.1016/j.ejps.2023.106401] [PMID: 36750147]

[93]   Ding X, Li S, Tian M, *et al.* Facile preparation of a novel nanoemulsion based hyaluronic acid hydrogel loading with *Poria cocos* triterpenoids extract for wound dressing. Int J Biol Macromol 2023; 226: 1490-9.
[http://dx.doi.org/10.1016/j.ijbiomac.2022.11.261] [PMID: 36442559]

CHAPTER 7

Antimicrobial Effect of Natural Products against Bacteria, Fungi, and Yeasts

Mai M. Badr[1,*] and **Israel Valencia Quiroz**[2]

[1] *Department of Environmental Health, High Institute of Public Health (HIPH), Alexandria University, Alexandria, Egypt*

[2] *Phytochemistry Laboratory, UBIPRO, Superior Studies Faculty (FES)-Iztacala, National Autonomous University of Mexico (UNAM), Tlalnepantla de Baz, México State, 54090, México*

Abstract: Antibiotics are compounds that either halt or destroy bacterial growth. They may be natural, semi-synthetic, or synthetic. Secondary metabolites, such as those produced by plants, animals, and microorganisms, are known as natural antimicrobials. The antibacterial/antimicrobial properties of secondary metabolites have been investigated over the past 30 years. Compounds derived from plants and culinary seasonings, including essential oils (EOs), are widely utilized in the food industry as organic agents to inhibit microbial growth in foods and prolong the shelf life of food products. Animal peptides (*i.e.*, polypeptides) also exhibit antimicrobial properties. Certain pathogenic and decaying bacteria may be inhibited by various chemicals produced by numerous microorganisms. Most microbially-derived antibacterial compounds are produced as intermediate byproducts of food fermentation. Numerous factors influence the antibacterial efficacy potential of natural products, including the source of the biological agent, harvesting time, the stage at which it is cultivated, and production methods.

Keywords: Animal origin, Antimicrobial origin, Byproducts, Essential oils, Natural products, Plant origin, Secondary metabolites.

INTRODUCTION

Initially hailed as "miracle medications," antibiotics have quickly become overused due to their widespread use. Over the past decade, pathogenic microorganisms have developed resistance to several antibiotic medications, which is difficult to overcome. To address this, pharmaceutical companies are attempting to create new antibiotics [1].

---

[*] **Corresponding author Mai M. Badr:** Department of Environmental Health, High Institute of Public Health (HIPH), Alexandria University, Alexandria, Egypt; Tel: +201129876935; E-mail: mai.badr@alexu.edu.eg

Natural products refer to chemicals that are produced by organisms or are acquired from the environment and have pharmacological or biological activity [2]. Primary and secondary metabolites produced by living things have diverse biological roles. Primary metabolites exert vital functions within the organism; by contrast, secondary metabolites may be superfluous byproducts or they may confer a substantial benefit to the organism. Secondary metabolites can be advantageous for human use and are used to treat several conditions, including cancer, bacterial infections, and inflammation, among others [3], with many compounds identified as having antibacterial properties. The majority of secondary metabolites with antimicrobial action originate from [1] plants such as fruits, vegetables, seeds, herbs, and spices [2], animal-based products such as milk, eggs, and tissue, and [3] microbes such as bacteria and fungi [4, 5].

Approximately 30,000 antimicrobial molecules have been identified in plants, and over 1340 species have known antibacterial properties [4]. Numerous naturally occurring antimicrobial compounds are derived from a diverse range of organisms, such as plants, animals, minerals, bacteria, and other distinct species [6]. Botanicals, which are commonly referred to as herbal products, are composed of an assortment of diverse plant components, such as seeds, bark, stems, leaves, flowers, fruits, wood, and even the whole plant itself [7]. A variety of animal-derived products exist, including beef, milk, eggs, meat, marine fish, sponges, curd, cow urine, cow dung, and other marine organisms [8]. In addition, natural products can be derived from honey and other sweeteners. They can also come from minerals, bacteria, and various other sources. The aim of the scientific field known as ethno-pharmacology, which explores substances such as natural antibiotics, is to apply the extensive pool of knowledge accumulated by indigenous individuals from specific communities and regions in relation to the numerous flora and fauna they have utilized to protect their well-being [8].

The term "antimicrobial activity" pertains to the ability to kill or inhibit microorganisms, thus halting their associated ailments. Various substances, such as alkaloids, tannins, terpenoids, essential oils, flavonoids, lectins, proteins and polypeptides, quinones, coumarins, polyphenols and phenolic compounds, enzymes, lysozymes, phagocytic cells, and numerous other organic constituents, possess active principal or secondary metabolites (or in the case of plants, phytochemicals) that exhibit antimicrobial properties [8]. Interest in natural antimicrobials has been reignited due to elevated public awareness surrounding the limited efficacy of synthetic pharmaceutical products in the management of numerous infectious diseases, which is attributed to drug resistance that has arisen from the incorrect usage and overprescription of antibiotic medications [8]. Additionally, several synthetic compounds can be harmful and are associated with unintended consequences. Increasingly, individuals are seeking better alternatives in order to maintain control over their medical treatment and avoid unnecessary toxins and chemicals [8].

Natural antimicrobial agents appear to be a feasible solution to address the numerous issues associated with increasing antibiotic resistance. Compared to synthetic antimicrobial agents, which are formed *via* amalgamations of chemicals and other synthetic methodologies [9, 10], there is a significant demand for new kinds of potent and healthful antibacterial chemicals that could avoid contamination in food and protect consumers from illness. This chapter aims to provide an overview of recent research on naturally occurring antimicrobial compounds derived from plants, animals, and microorganisms. The primary objective is to investigate the feasibility of using these compounds as a potential treatment option for microbial pathogens responsible for causing various human diseases [10].

EXPLORING THE ORIGINS AND DIVERSITY OF NATURAL ANTIBIOTICS

Botanical Origins: Natural Antimicrobial Substances

Since ancient times, humans have been aware of the healing powers of plants. Throughout history, the knowledge of utilizing plants and their byproducts for medicinal purposes has been passed down over generations in diverse regions of the world [11]. This has greatly influenced the growth of diverse traditional medical systems. According to the World Health Organization (WHO), nearly 80% of people worldwide regularly utilize traditional herbal remedies [12].

Plants generate a vast array of diverse chemical compounds, several of which do not have indispensable requisite functions in fundamental metabolic processes, but represent adaptation of the plant's adverse biotic and abiotic conditions [13]. Such chemicals that are categorized as substances with physiological activity are sometimes referred to as secondary metabolites since they are produced as byproducts or intermediates of secondary plant metabolism. These ancillary compounds proficiently exert their role within the plant system and are associated with biological and pharmacological effects in humans in addition to defining distinctive botanical attributes such as the hue and fragrance of flowers and fruits, the zestful flavor of spices, and the savory taste of vegetables [11]. Consequently, the medicinal properties of plants are associated with plant secondary metabolites [14].

Extracts of culinary, therapeutic, and herbal plants are used to create essential oils (EOs), which are also products of secondary metabolism [15]. Antibiotics produced by plants can be divided into two categories: phytoanticipins, which are preformed compounds possessing inhibitory properties, and phytoalexins, which are synthesized from intermediates through novel pathways in response to microbial invasion [11, 16].

Gram-positive bacteria are commonly more susceptible to inhibition by active ingredients from plants than their gram-negative counterparts [17]. The presence of widespread metabolic toxins or a wide variety of antibiotic compounds may explain their effectiveness against both types of bacteria [18]. Phenolics and polyphenols are two major categories of secondary metabolites with antibacterial activity. This collection of substances includes phenols, phenolic acids, alkaloids, quinones, flavones, flavonoids, flavonols, tannins, and coumarins as second-generation antibacterial metabolites. Phenols, a chemical class of molecules, are formed by an aromatic phenolic group bonding with a hydroxyl functional group (-OH) [19]. The toxic potential of phenolic compounds toward microorganisms is influenced by the number and location of hydroxyl groups within the structure [20]. Currently, available evidence suggests that augmenting the hydroxylation process can enhance this toxic effect [21, 22].

Alkaloids, phenolics, and terpenes are the three major chemical families that antibacterial secondary metabolites are typically categorized into. The antibacterial efficacy of plant constituents is contingent upon the chemical composition of the active constituents and their respective concentrations. Plants contain a multitude of chemical components, including saponins, flavonoids, glucosinolates, phenolic compounds, thiosulfinates, and organic acids, which display noteworthy antimicrobial properties [10]. Nevertheless, it is noteworthy that phenolic compounds, including terpenes, aliphatic alcohols, aldehydes, ketones, acids, and isoflavonoids, serve as the primary constituents in plants that exhibit antibacterial properties [15, 23].

Quinones possess two ketone substitutions in their aromatic rings. Quinones may irreversibly bind with the nucleophilic amino acids found in proteins, which may explain their antimicrobial qualities. Targets in the microbial cell include membrane-bound enzymes, cell wall polypeptides, and exposed adhesins on the cell surface. Phenolic compounds known as flavones have a single carbonyl group and the addition of a 3-hydroxyl group results in the formation of a flavonol [11]. Flavonoids, despite being present as a C_6-C_3 unit linked to an aromatic ring, exhibit hydroxylated phenolic characteristics [24]. Given that plants produce flavones, flavonoids, and flavonols in reaction to microbial infection, it is not surprising that these compounds are efficient antimicrobials *in vitro* against various pathogens [22].

The probable efficacy of polyphenols stems from their capability to engage with extracellular and soluble proteins, alongside bacterial cell walls and various other substances [25]. The disruption of membranes by lipophilic flavonoids is another possibility [22]. Tannins comprise a category of polymeric phenolic constituents that are ubiquitously found in virtually all plant components [26]. The components of the plant, namely, the bark, wood, leaves, fruits, and roots, are classified into two distinct groups: condensed and hydrolysable tannins [27]. During microbial attacks, tannins are produced and stored in plant tissue. The potential antibacterial efficacy of these compounds may be attributed to their capacity to inactivate microbial adhesives, enzymes, and cell membrane protein carriers, owing to a characteristic property commonly referred to as astringency. In the group of coumarins, which are benzo-pyrones, two distinct types can be identified: simple and cyclic coumarins, namely, furanocoumarins and pyranocoumarins [11, 28]. Coumarins have been observed to stimulate macrophages, which may indirectly worsen infections [22].

Terpenoids represent the oxygenated counterparts of isoprene-based terpenes and belong to a vast and heterogeneous chemical compound class [29]. Terpenes are classified based on the number of isoprene subunits present in their chemical structure. There are several subclasses of terpenes, including monoterpenes (C_{10}), sesquiterpenes (C_{15}), diterpenes (C_{20}), triterpenes (C_{30}), tetraterpenes (C_{40}), and polyterpenes [11]. Alkaloids, which are classified as nitrogen heterocycles, rank among the first bioactive substances of plant origin that were identified [11]. Since they are composed of amino acids, nitrogen confers these compounds with alkaline characteristics. The proposed mechanisms underlying the antibacterial effect of these agents include their capacity to integrate with DNA, the blockade of various enzymes such as esterase, DNA polymerase, and RNA polymerase, and a reduction in cellular respiration [11, 30].

Phenolic constituents were proposed to be responsible for the antibacterial properties of 46 spice and herb extracts [31]. The authors of this study found that particular spices under investigation strongly inhibited *Bacillus cereus*, *Listeria monocytogenes*, *Staphylococcus aureus*, *Escherichia coli*, and *Salmonella anatum* [10]. The antibacterial properties of red cabbage are associated with anthocyanin polyphenols, which are abundant in the plant material [32].

Phenolic constituents in essential oils, such as those found in citrus oils such as bergamot, orange, and lemon, as well as in tea tree oil (terpenoids) and olive oil (oleuropein), exhibit broad-spectrum antibacterial properties. The widespread knowledge concerning the efficacy of nonphenolic oil components derived from oregano, clove, cinnamon, citral, garlic, coriander, rosemary, parsley, lemongrass, muscadine seeds, and sage against both gram-positive and gram-negative groups continues to increase [10, 33, 34]. Allyl isothiocyanate and garlic oil, both of which are nonphenolic constituents of essential oils, are highly effective in combating gram-negative bacteria [35]. Numerous botanicals sourced from herbs, spices, fruits, and vegetables have antimicrobial activity, including guava, xoconostle, pepper, cabbage, and onion, in addition to seeds and leaves, such as grape seeds, caraway, fennel, nutmeg, parsley, and olive leaves [10, 15].

The Utilization of Herbs and Spices for Antimicrobial Applications

Humans have employed the use of herbs and spices for various reasons, including antibacterial agents, since the dawn of time. Natural antimicrobials that are frequently utilized include extracts derived from herbs, spices, and other plant sources [36]. They typically include EOs, which are widely recognized to have antibacterial activity. The vapor pressure of EOs from plants is often quite high, which allows Eos to penetrate the liquid and gas phases to reach microbial pathogens [37].

The EOs that are most efficacious for the inhibition of *S. enteritidis*, *E. coli*, and *Listeria innocua* growth are lemongrass, cinnamon, and geraniol [10, 38]. Both *E. coli* O157:H7 and *S. typhi* are observed to be susceptible to mustard essential oils [39, 40]. A series of experiments evaluating the antibacterial properties of several species against *Salmonella* and other enterobacteria found that spices exhibit varying levels of effectiveness. For example, clove demonstrated superior antibacterial efficacy, followed by kaffir lime rind, cumin seeds, cardamom pods, coriander seeds, nutmeg, mace spice, root of ginger, garlic cloves, holy basil leaves, and finally, the foliage of the kaffir lime [3, 41, 42]. Due to the potent bacteriostatic and bactericidal properties of EOs derived from oregano and thyme, *E. coli* O157:H7 cells were irreparably damaged upon exposure to specific bactericidal levels of oregano essential oils within a mere minute [43]. Eos from oregano and thyme were observed to exhibit greater antibacterial efficacy compared to those of clove and bay [31, 44].

Harnessing the Antibacterial Potentials of Fruits and Vegetables

Fruits and vegetables have received greater attention recently for their association with antibacterial action against various pathogenic and spoilage microbes that are attributed to the phenolics and organic acids found within. For example, the antibacterial properties of capsicum are associated with phenolic substances and 3-hydroxycinnamic acid, also known as coumaric acid [10, 45]. Flavonoids extracted from the processing of bergamot peel exhibit antibacterial properties against gram-negative bacteria such as *E. coli*, *P. putida*, and *S. enterica,* and this antibacterial activity is augmented by enzymatic deglycosylation [46 - 48].

A wide range of fruits and vegetables, namely, asparagus, bell peppers, beets, carrots, cucumbers, red onions, red cabbage, rhubarb, rutabaga, raspberry, pomegranate, spinach, strawberries, and green tea, have been investigated for their potential antibacterial properties [46, 49, 50]. Scientists have discovered that juices derived from purple and red fruits and vegetables possess antibacterial properties that combat *S. epidermidis* and *Klebsiella pneumoniae*. However, juices extracted from all green vegetables do not possess similar antibacterial properties. It has been noted that pomegranate can inhibit *E. coli* O157:H7 [51, 52]. A methanolic extract of pomegranate peel exhibited antibacterial properties against both gram-positive and gram-negative bacteria as well as fungi. This can be attributed to the presence of phenolics and flavonoids within the extract [53].

Another study demonstrated that guava extract inhibited the growth of *Salmonella spp.* and *E. coli* O157:H7 [54, 55], and sustainably sourced crude extracts of xoconostle pears exhibited robust antibacterial properties against *E. coli* O157:H7 [56]. In another study, the antibacterial efficacy of quince (*Cydonia oblonga* Miller) pulp and peel liquid extract, obtained through acetone, was shown to be greater than that of quince pulp, owing to its significantly higher concentration of phenolic compounds, specifically chlorogenic acid. This finding highlights the potential of quince fruit as a rich source of natural antimicrobial agents [10, 57].

Garlic extract (*Allium sativum*) is known to possess a diverse range of antibacterial properties. Allicin, an organosulfur compound present in garlic, is known to hinder the growth of both gram-positive and gram-negative bacteria belonging to various genera, including *Escherichia*, *Salmonella*, *Streptococcus*, *Staphylococcus*, *Klebsiella*, *Proteus*, and *Helicobacter* [3, 58]. Allicin, diallyl thiosulfinic acid, and diallyl disulfide are all potentially significant factors contributing to the antibacterial characteristics of garlic [3, 59]. Moreover, compared to phenolic components in garlic, organosulfur compounds may exhibit stronger antibacterial action [3].

Essential oils derived from black pepper, clove, geranium, nutmeg, oregano, and thyme exhibit antibacterial properties against 25 diverse genera of bacteria, with varying degrees of growth inhibition [10, 60]. The methanolic extracts of Ghuiya (*Colocasia esculenta*), Suran (*Amorphophallus campanulatus*), Pumpkin (*Cucurbita pepo*), and Spinach (*Spinacia oleracea*) have demonstrated varying levels of antibacterial activity ranging from moderate to significant against a diverse range of gram-positive and gram-negative bacteria [61, 62]. Methanolic extracts of spinach were shown to outperform aqueous extracts, inhibiting *E. coli* more effectively. Studies have shown that red cabbage possesses the potential to combat fungi associated with human fungal illnesses, specifically *Trichophyton rubrum* and *Aspergillus terreus*, in addition to the antibiotic-resistant strains such as methicillin-resistant *S. aureus*, *E. coli* O157:H7, *Pseudomonas aeruginosa*, *Klebsiella pneumoniae*, *S. aureus*, and *S. enterica* serovar *Typhimurium* [10, 32]. Fruits and vegetables that are abundant in organo- and phenolic acids are therefore more likely to possess strong antibacterial action.

Antimicrobial Properties in Seeds and Foliage

Essential oils, seeds, and leaves that are rich in phenol possess antibacterial properties. Notably, crude essential oils of eucalyptus (*Eucalyptus dives*), coriander (*Coriandrum sativum L.* seeds), cilantro (immature *C. sativum L.* leaves), and dill (*Anethum graveolens L.*) have been found to be highly effective against six types of bacteria and one type of yeast at concentrations below 0.5% (vol/vol) [10, 63]. Olive leaves also are effective against harmful bacteria and fungi. Aqueous extracts of olive leaves at 0.6% (w/v) were observed to destroy a significant proportion of bacterial cells within a three-hour time frame [64], although it is imperative to note that dermatophytes and *Candida albicans* necessitate 1.25% and 15% (w/v) extracts, respectively [46, 65].

Additionally, powerful antibacterial and antifungal capabilities were observed in the olive leaf extract. The antibacterial and antifungal properties of olive leaves are attributed to phenolic compounds such as caffeic acid, verbascoside, oleuropein, luteolin 7-O-glucoside, rutin, apigenin 7-O-glucoside, and luteolin 4'-O-glucoside [10, 66]. Therefore, utilization of olive leaf extract, specifically as a potential source of phenolic compounds, is an area of focus in the nutraceutical industry [67].

Methanolic extracts of the flower, stem, and leaf of *Tagetes minuta* L. exhibited significant antibacterial activity against a range of bacterial strains, including *E. coli*, *Klebsiella pneumoniae*, *Pseudomonas aeruginosa*, *S. typhi*, *S. aureus*, *Streptococcus viridians*, *B. licheniformis*, *B. subtilis*, and *Pasteurella multocida*, with MICs in the range of 4-100 mg/mL [10, 68]. The authors hypothesized that the existence of alkaloids, tannins, saponins, and flavonoids might contribute to the observed antibacterial effects. Additionally, the antibacterial effects of *Rubus chamaemorus* leaves were assessed against *Candida albicans*, along with gram-positive and gram-negative microorganisms [10, 68, 69].

Caffeic acid, chlorogenic acid, and protocatechnic acid are phenolic chemicals found in coffee that have been discovered to possess antimicrobial properties [10, 70]. The methylated xanthine alkaloid derivative caffeine (1,3,7-trimethylexanthine) is found in various plant species. At a concentration of 0.5%, the inherent caffeine present in coffee significantly impeded the proliferation of *E. coli* O157:H7 [71]. Similarly, the polyphenols present in tea, such as epicatechin, catechin, caffeine, chlorogenic acid, gallic acid, theobromine, theophylline, gallocatechin, epigallocatechin gallate, catechin gallate, epicatechin gallate, and theaflavin, exhibit antimicrobial properties against both gram-positive and gram-negative bacteria [10]. *Puerariae radix*, known for its water-soluble crude arrowroot tea, exhibits remarkable antibacterial activity against *S. aureus*, *L. monocytogenes*, *E. coli* O157:H7, and *S. enterica* [72, 73].

The antibacterial properties of arrowroot tea may be due to its catechin content. Epigallocatechin gallate and epigallocatechin are two catechins that have been identified in green tea extract and their molecular structure, which includes a galloyl moiety, has been found to confer potent antibacterial activity [41, 74].

Antimicrobial Substances Sourced from Animals

Animals possess many antimicrobial defense mechanisms that are frequently developed as part of host defense mechanisms. Antimicrobial peptides, which are structurally polypeptides, contribute substantially to the innate antibacterial activity of these agents in mammals. These peptides are plentiful in the natural world and serve as integral components of overall defense mechanisms in hosts [10, 75]. Since they do require specific molecular binding sites and can rupture membranes rapidly preventing even fast-growing bacteria from mutating, these peptides may represent an antidote to antibiotic resistance [15]. Peptides, for the most part, exhibit antimicrobial activity that targets both gram-positive and gram-negative bacteria. In addition, they possess antifungal and antiviral characteristics [16, 18, 76].

Various foodborne pathogens, including *Vibrio parahemolyticus*, *Listeria monocytogenes*, *Escherichia coli* O157:H7, *Saccharomyces cerevisiae*, and *Penicillium expansum*, are susceptible to the peptide pleurocidin [77, 78]. However, the efficacy of pleurocidin is diminished with high levels of magnesium and calcium. Pleurocidin, which is derived from the epidermal mucous membrane of the winter flounder (*Pleuronectes americanus*), inhibits both gram-positive and gram-negative bacteria [15, 79]. Defensins, which belong to a distinct category of antimicrobial peptides, are found within the mammalian epithelial cells of chickens and turkeys and are ubiquitously distributed in nature. Their antimicrobial efficacy extends to a broad spectrum of microorganisms, including gram-positive and gram-negative bacteria, fungi, and viruses [10, 15].

Protamine and magainin are among other antimicrobial peptides that exhibit antimicrobial activity against both gram-positive and gram-negative bacteria, yeasts, and molds [10, 46, 80]. The capacity of gram-positive microorganisms to attach themselves to surfaces is diminished by magainin, which then destroys them [77, 81]. Nevertheless, dietary ingredients may interfere with the antibacterial actions of protamine, although this unwanted interference can be decreased by altering the electrostatic characteristics of protamine [81].

It is also important to note that milk possesses inherent antimicrobial peptides. For example, the copious lactoperoxidase enzyme found in milk exhibits potent antibacterial properties against a range of microorganisms, such as bacteria, fungi, and viruses. Notably, lactoperoxidase can be found in various types of milk, including cow, ewe, goat, buffalo, pig, and human milk [46, 82]. The lactoperoxidase enzyme present in human milk has the potential to aid the immune system in combatting preexisting infections located in the oral cavity and upper gastrointestinal tract [10, 83].

In addition to peptides, other studies have reported that certain animal fats and polysaccharides exhibit antibacterial properties. For example, chitosan, a naturally occurring linear polysaccharide derived from the exoskeletons of crustaceans and other arthropods, has been utilized as an active ingredient for its antifungal [84, 85] and antibacterial properties and is renowned for its special polycationic nature. The antibacterial activity of chitosan has been observed against both *S. aureus* and *E. coli*. It has been suggested that the molecular weight of chitosan plays a pivotal role in determining this activity [86, 87].

ANTIMICROBIAL COMPOUNDS ORIGINATING FROM MICROORGANISMS

Microbes, such as bacteria, fungi, and molds, are known to produce a variety of substances with antimicrobial properties. These substances, also known as secondary metabolites, include antibiotics such as penicillins, cephalosporins, tetracyclines, aminoglycosides, chloramphenicol, and macrolides [10, 88]. In addition to antibiotics, antimicrobial metabolites derived from microorganisms encompass bacteriocins, organic acids, hydrogen peroxide, ethanol, and diacetyl [89].

Low concentrations of organic acids can be beneficial without compromising the desired sensory qualities of food products. According to reports, there is evidence to suggest that organic acids demonstrate efficacy against bacteria in a specific sequence, namely, lactic acid, formic acid, acetic acid, and propionic acid. Additionally, it has been observed that organic acids exhibit greater effectiveness against gram-positive bacteria when compared to gram-negative bacteria [90]. For instance, compared to *E. coli* and *Salmonella sp.*, which are both gram-negative bacteria, *Clostridium perfringens* (a gram-positive bacterium) was found to be more susceptible to organic acids, as demonstrated through cultures [91].

A more pronounced inhibitory effect can be achieved by combining two organic acids or an organic acid with other naturally occurring antimicrobial compounds compared to the use of a singular organic acid, such as in the case of acetic and formic acids, lactic and formic acids, and propionic and formic acids, rather than each acid used in isolation [10, 92].

The combined action of ascorbic acid and lactic acid has been observed to impede the proliferation of *E. coli* O157:H7 [54, 93]. It has been similarly documented that *Salmonella* and *E. coli* O157:H7 demonstrate greater tolerance to lactic acid when used in conjunction with copper sulfate compared to its utilization in isolation [94, 95]. Owing to its strong oxidizing influence on bacterial cells, coupled with its capacity to disrupt the fundamental molecular structures of cell proteins, hydrogen peroxide exhibits pronounced antibacterial efficacy [46, 96]. Hydrogen peroxide, the primary metabolic byproduct of lactic acid bacteria (LAB), is posited to possess anti-infective attributes, particularly in instances of bacterial vaginosis [97, 98]. Diacetyl, also known as 2,3-butanedione, is another antibacterial compound produced by heterofermentative lactic acid bacteria (LAB), and its production occurs during the process of fermentation [46, 99].

Bacteriocins are proteinaceous antibacterial substances that are frequently produced by various species of LAB. They constitute a subclass of heterogeneous antimicrobial peptides synthesized *via* ribosomal pathways [94, 100]. Notably, bacteriocins are produced by both gram-positive and gram-negative bacteria [56, 101]. Bacteriocins, recognized as cationic peptides bearing amphiphilic or hydrophobic traits, primarily target the bacterial membrane. They are divided into three distinct categories: Class I encompasses lantibiotics; Class II comprises small, heat-stable nonlantibiotics; and Class III, includes larger, heat-labile bacteriocins [100, 102].

Among the most notable bacteriocins are nisin, diplococcin, acidophilin, pediocin, bulgarican, helveticin, lactacin, and plantaricin [101, 103]. A modified milk medium is fermented by LAB species, primarily *Lactococcus lactis*, to create nisin; however, low pH levels are required to achieve this [104]. *P. acidilactici* and *P. pentosaceus* generate a bacteriocin known as pediocin. Pediocin is a bacteriocin that is typically used in fermented sausage and is produced by the bacterium *Pediococcus acidilactici* and *P. pentosaceus*. Pediocins, recognized as thermostable proteins, demonstrate robust bactericidal activity against gram-positive bacteria, which include food spoilage and pathogenic species such as *L. monocytogenes*, *Enterococcus faecalis*, *S. aureus*, and *C. perfringens* [10, 105]. Natamycin, another bacteriocin produced by *Streptomyces natalensis*, is a polyene antifungal that works well against molds and yeasts but not bacteria or viruses

[106]. The solubility of natamycin in water is exceedingly poor, and works only at very low concentrations.

Reuterin and reutericyclin, potent antimicrobial compounds produced by *Lactobacillus reuteri*, exhibit strong activity against gram-positive bacteria. Furthermore, reuterin demonstrates antibacterial activity against yeasts, molds, protozoa, and gram-negative bacteria [107, 108]. Reuterin is a compound that exists in a mixture of different forms: monomeric, hydrated monomeric, and cyclic dimeric variants of β-hydroxypropionaldehyde [21, 109]. Reutericyclin, a tetramic acid derivative, and reuterin, a nonproteinaceous, water-soluble metabolite of glycerol, also known as β-hydroxypropionaldehyde, are two notable substances [110]. Various studies suggest that reuterin demonstrates considerable antimicrobial action against a variety of bacterial strains. These include *L. monocytogenes*, *E. coli* O157:H7, *S. choleraesuis subsp. choleraesuis*, *Yersinia enterocolitica*, *Aeromonas hydrophila subsp. hydrophila*, and *Campylobacter jejuni* [10, 111].

Categorizing Natural Compounds with Antimicrobial Properties

Natural antimicrobials encompass a broad range of agents with antibacterial, antifungal, antiviral, antiprotozoal, and insecticidal properties. In the realm of antibacterial agents, the leaves of *Cassia alata* have demonstrated *in vitro* effectiveness against a variety of microorganisms, including *Salmonella typhi*, *E. coli*, and *Staphylococcus aureus* [8, 112]. Extracts of the dried nuts of *Semecarpus anacardium* have demonstrated bactericidal efficacy against the gram-positive pathogens *Staphylococcus aureus* and *Corynebacterium diptheriae* [113], and are similarly effective against three gram-negative strains, namely, *E. coli*, *S. typhi*, and *Proteus vulgaris* [114, 115]. Essential oils from thyme, cinnamon, and eucalyptus have all demonstrated antibacterial properties [116].

Among antifungal agents, leaves from the neem tree display anti-dermatophytic action and, thus, are effective as antifungals [117]. Various plants exhibit antifungal activity to varying extents, including eucalyptus (88%), tulsi (85.5%), neem (84.66%), castor (75%), and jatropha (10%) [8, 118]. The essential oils from the leaves of *Aegle marmelos* (Bael) had 100% sporicidal potency and antifungal activity [119]. The leaves of *Cassia alata* have demonstrated antifungal effectiveness against *Candida albicans* and *Aspergillus niger* [72, 120], and the roots of *Withania somnifera*, or Aswagandha, are effective against *Aspergillus fumigatus* [121].

Among antiviral agents, at high concentrations, the triterpenoid glycyrrhizin, which is found in licorice (*Glycyrrhiza glabra*), was effective against RNA viruses such as measles virus, poliovirus types 1, 2, and 3, and HIV and was able to inhibit DNA viruses at lower concentrations [122]. Certain extracts from mangrove plants have demonstrated antiviral efficacy against the HIV virus, and function by preventing the virus from binding to host cells [8, 123]. Extracts from the leaves of *Hemidesmus indicus* and the stems of *Cassia fistula* stop the Ranikhet Disease (RD) virus from reproducing and elicit cytopathic effects as well as the vaccinia virus due to their interferon-like components [124].

Quinine, a medication used in antiprotozoal drugs and known for its antimalarial properties, is derived from the bark of the cinchona tree [8, 118]. Extracts of *Artemisia japonica* have been shown to suppress the schizont stage in chloroquine-sensitive strains of *Plasmodium falciparum* [125]. *Swertia chirata* extracts exhibited anti-leishmanial action and inhibited the action of the topoisomerase-I enzyme as a catalyst from *Leishmania donovoni* [126]. Trypanocidal activity was shown by *Parthenium hysterophorus* (congress grass) extracts against *Trypanosoma evansi* [127].

Understanding the Mechanisms of Action of Natural Antibacterial Agents

The primary antibacterial components in plants are phenolic compounds. The bactericidal effects of phenolic compounds are widely established [128], although the precise antimicrobial mechanism is unclear. The antibacterial properties of phenolic compounds may be attributed to their ability to alter the susceptibility of microbial cells, leading to the loss of macromolecules from within the cell [129]. Another possibility is that phenolic chemicals interact with membrane proteins and impair membrane function, leading to changes in membrane structure and functionality (Fig. **1**) [130]. When phenolic compounds are combined, they can have synergistic antibacterial actions and improve the antimicrobial response compared to the reaction of each component alone [131].

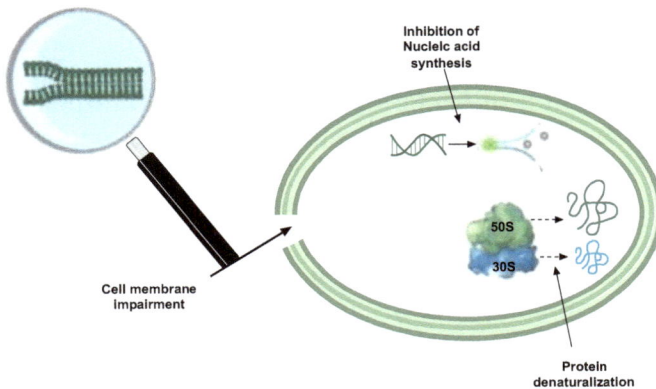

Fig. (1). Primary mechanism of action of antibiotics.

The impact of phenolic substances might vary depending on concentration. For example, phenols breakdown proteins at high concentrations, but impact enzyme activity only at low concentrations. The antibacterial activity of isothiocyanates derived from onion and garlic are thought to inactivate extracellular enzymes through the oxidative cleavage of disulfide bonds. The generation of the active thiocyanate radical has been posited as a mediator of this antimicrobial effect, as illustrated in Fig. (**1**) [10, 131].

Antimicrobial peptides appear to have a multitargeted mechanism of action, which is typical for peptides. Despite the potential *via*bility of internal targets, the plasma membrane continues to be the most frequently identified target of peptides [10, 132]. Most antimicrobial peptides act through nonspecific mechanisms but in some cases, selectivity is exhibited with certain pathogens. Antimicrobial peptides can have amphipathic properties that permit them to interact directly with the membrane of the microorganism. This action quickly results in membrane disruption at multiple sites, which causes critical cell components to leach out [80].

The antibacterial activity of individual fatty acids is influenced by its shape and structure, specifically the number of double bonds and the length of the carbon chain [133]. Longer carbon chain fatty acids typically have a larger inhibitory effect than shorter chain fatty acids. Moreover, unrefined fatty acids typically have a stronger inhibitory impact than refined fatty acids, and medium- and long-chain unsaturated fatty acids are more effective against gram-positive bacteria than gram-negative bacteria [134]. The antibacterial activity of saturated fatty acids decreases chain length lengthens or shortens [10, 135]. Fatty acids displaying the highest activity typically have a chain comprising 10 or 12 carbons [10, 136]. Generally, lipids exert their antibacterial actions either by disrupting the bacterial cell wall or membrane, preventing intracellular replication, or inhibiting an intracellular target, thereby rendering the microorganisms inactive.

The antimicrobial action of bacteriocins results in the formation of pores in the cytoplasmic membrane of target bacteria [15]. Consequently, small intracellular molecules and ions are lost, leading to the dissipation of the proton motive force [94, 137]. This process of pore formation in the cytoplasmic membrane explains why nisin is less effective against gram-negative bacteria, as the outer membrane hinders the entry of this molecule to its site of action [15, 138]. In alignment with the mode of action of nisin, the substance first permeates through the cell wall of gram-positive bacteria. Nonetheless, depending on its composition, thickness, or hydrophobicity, the gram-positive cell wall can act as a molecular sieve for nisin [10, 139].

After bypassing the cell wall nisin attaches to the cytoplasmic membrane of the target microorganism. The phosphate groups of the surface membrane phospholipids, which carry negative charges, will then interact electrostatically with nisin. Nisin disrupts the functioning of the bacterial cytoplasmic membrane by interacting with phospholipids in the membrane. This prevents spore germination, thereby inhibiting the formation of spores [35]. Within the cheese industry, nisin is used to prevent the formation of *Clostridium spp.* since it is particularly effective against various gram-positive bacteria [140].

Influential Factors in the Antibacterial Activity of Natural Products

The nature of natural substances, their structure, and functional groups, as well as their botanical source, harvesting period, development stage, and extraction technique, may have an impact on antibacterial activity [15]. The effectiveness of natural antimicrobials is also significantly impacted by pH. For example, improvements in the antiseptic activity of EOs against *L. monocytogenes* appear are observed to be strongly influenced by low pH values (approximately 5) [141]. At a lower pH, EOs may become more hydrophobic, which enables them to dissolve more easily in the lipid component of the target bacteria's cell membrane.

Methods of Extracting Natural Antibacterial Compounds

The most popular methods for producing plant-based antimicrobials on a commercial scale are steam distillation (SD) and hydrodistillation (HD), although alternative techniques, such as supercritical fluid extraction (SFE), provide rapid mass transfer rates and improved solubility. However, when a specific compound is sought, varying factors such as temperature and pressure influence whether additional components are extracted, leading to a less pure extract [10]. Bioengineering of EO components also increases the number of commercially available products [142]. The antibacterial properties of plants are primarily attributed to EOs and other extracts from plants. Many techniques, including steam, cold, dry, and vacuum distillation, can be used to extract these compounds from plants and spices [132].

Typical extraction techniques involve heat or chemical treatments that may change the functioning or natural properties of the active components or result in the production of hazardous substances [143]. The direct extraction method can be employed to acquire extracts from fruits and vegetables [56]. The native structure of the active components is neither altered nor destroyed during this straightforward and safe process. Moreover, direct extracts, such as that from guava and xoconostle pears, are demonstrated as effective against *Salmonella spp.* and *E. coli* O157:H7, respectively [10, 56, 74]. Mechanical methods can be used to extract fruits and vegetables directly, without the need for chemicals, heating, or concentration techniques. Additionally, direct extracts can be safely added to culinary products.

Water extraction is another simple method that can be used to extract water-soluble phenolic compounds from seeds and leaves [10]. Studies have shown that heat treatment applied to water-soluble muscadine seed extracts enhances acidity, total phenolics, specific phenolic compounds, and antibacterial activity [144, 145]. Other methods, including ultra-high-temperature processing, far-infrared radiation, and/or enzymatic procedures, may enhance the dispersion of low-molecular-weight compounds from the polymer structure, thereby diminishing antibacterial activity [10, 146]. Consequently, direct extraction or water extraction appears to be the most efficient approach to prevent potential modification or alteration of the nature of organic antibacterial compounds [10].

Increasing Resistance to Natural Products

There are several mechanisms through which a bacterium can counteract the effects of an antimicrobial molecule. These include target modifications, expulsion or modification of the antibiotic, inactivation, reduced permeability, and biofilm formation [147, 148]. These resistance mechanisms, as illustrated in Fig. (**2**), may arise naturally through mutations, or they can be acquired *via* processes such as transduction, transfection, or conjugation [147]. To develop long-lasting antimicrobials, it is crucial to understand how bacteria acquire resistance to antibiotics.

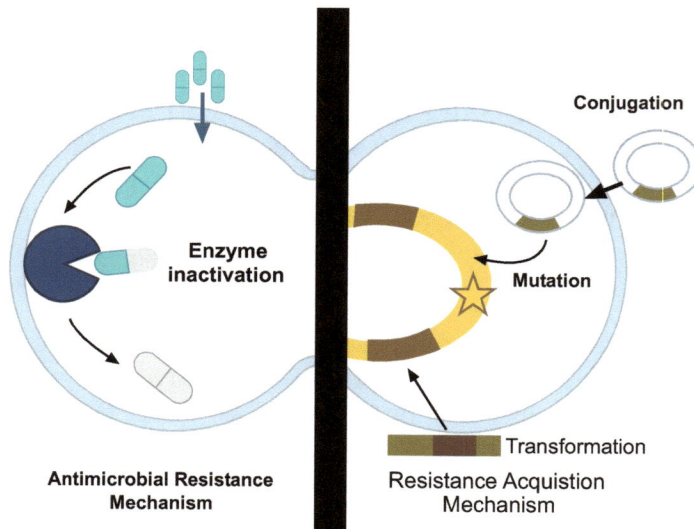

Fig. (2). Mechanisms of antimicrobial resistance and acquisition in bacteria.

Some natural products are more likely than others to encounter microbial resistance depending on individual characteristics. When bacteria interface with antimicrobial molecules that attack extremely common targets, it is less likely that alterations in these sites that would provide resistance to occur changes to one or more essential pathways or targets may result in an unacceptably high fitness cost for the bacteria [147, 149]. On the other hand, less conserved molecular targets are more likely to prompt the development of resistance when medications are utilized [147]. Bacteria have a greater capacity to incorporate changes to less evolutionarily conserved or nonessential targets, given that they can more freely modify the molecular target or adjust internal metabolic processes without substantial fitness costs. Despite the evolution of antimicrobial resistance mechanisms leading to reduced vigor in the absence of a selective environment, compensatory mutations or alterations in epistasis may alle*via*te this decrease in adaptive efficiency [147, 149].

To offset the impact of a multifactorial antimicrobial agent, a bacterium would need to experience numerous concurrent molecular alterations, which would adversely affect metabolism and impose a high fitness cost, which is incompatible with growth. Likewise, within bacterial populations, mutations that carry a high fitness cost are less likely to persist once the selective pressure is withdrawn [147, 150].

PUBLIC PERCEPTION AND CONCERNS REGARDING ADVERSE EFFECTS OF NATURAL ANTIMICROBIALS

Antibiotics are recommended to speed patients' recovery and to stop illnesses from spreading [151]. Each of these considerations is crucial to avoid toxicity in patients, as antibiotics, like all other categories of drugs, possess both generic and specific side effects [152, 153]. Adverse effects of antibiotics encompass hypersensitivity, hematological reactions, and impacts on the neurological, gastrointestinal, renal, cardiac, pulmonary, and hepatic systems [154, 155].

In comparison to synthetic antimicrobial agents, organic antibacterial substances produced from plants, animals, and microorganisms are thought to be safer. However, this depends on the concentrations and dosages of the drug's active components. For example, high concentrations of garlic are not advised for patients undergoing surgery or for those taking blood thinners since garlic may increase the risk of bleeding. Many phytochemicals are effective against bacteria, though commercial preparations marketed as antibiotics are not yet available. Therefore, it is important to investigate the therapeutic effects of plants to gain a deeper understanding of their safety and efficacy in order to promote the use of herbal remedies in place of and as an alternative to synthetic medications [152].

Antibiotics are administered to treat illnesses quickly and effectively while preventing the spread of highly contagious diseases. Antibiotics should only be taken as directed, prescriptions should not be shared, and the dosage prescribed should be taken completely before they expire, even when symptoms disappear. Because bacterial infections pose a considerable risk to human life, medical professionals are endeavoring to create herbal medications that can overcome antibacterial resistance [151]. Natural antimicrobial compounds from various sources can occasionally impart taste and aromas to food. Various components, such as proteins, lipids, complex carbohydrates, and sugars, found in our diet can potentially impede the antibacterial activity that is otherwise exerted by certain dietary constituents [155, 156].

CONCLUSION

Antibiotics, which are regarded as a cornerstone of contemporary medicine, are the primary approach taken for treating infectious diseases. However, new antimicrobial medicines are desperately needed, as existing antibiotics are rapidly losing their effectiveness. Since ancient times, people have understood that items existing naturally in the environment have antibacterial properties. Natural products play an important role in contemporary medicine due to their antibacterial properties. Antimicrobial substances are extremely valuable and can be sourced from organisms. This section provides an analysis of the antibacterial effects displayed by the most prominent natural substances sourced from animals, bacteria, fungi, and plants. Natural antimicrobials are widely acknowledged as safe and appear to be the best solution for overcoming resistance to antibiotics. This chapter provides an overview of the potency of a variety of natural products in combating microbes that have evolved drug immunity, the underlying mechanisms of action, the likelihood of resistance developing against these substances, their relevance in contemporary health care, and future directions. Moreover, concerns regarding the dosage and concentration of some active ingredients are discussed.

REFERENCES

[1]   Saleem M, Nazir M, Ali MS, *et al.* Antimicrobial natural products: An update on future antibioticdrug candidates. Nat Prod Rep 2010; 27(2): 238-54.
        [http://dx.doi.org/10.1039/B916096E] [PMID: 20111803]

[2]   Koehn FE, Carter GT. The evolving role of natural products in drug discovery. Nat Rev Drug Discov 2005; 4(3): 206-20.
        [http://dx.doi.org/10.1038/nrd1657] [PMID: 15729362]

[3]   Gyawali R, Ibrahim SA. Impact of plant derivatives on the growth of foodborne pathogens and the functionality of probiotics. Appl Microbiol Biotechnol 2012; 95(1): 29-45.
        [http://dx.doi.org/10.1007/s00253-012-4117-x] [PMID: 22622837]

[4]   Tajkarimi MM, Ibrahim SA, Cliver DO. Antimicrobial herb and spice compounds in food. Food

Control 2010; 21(9): 1199-218.
[http://dx.doi.org/10.1016/j.foodcont.2010.02.003]

[5]   Saeedi P, Petersohn I, Salpea P, *et al.* Global and regional diabetes prevalence estimates for 2019 and projections for 2030 and 2045: Results from the International Diabetes Federation Diabetes Atlas, 9th edition. Diabetes Res Clin Pract 2019; 157: 107843.
[http://dx.doi.org/10.1016/j.diabres.2019.107843] [PMID: 31518657]

[6]   Aljaloud SO, Gyawali R, Reddy MR, Ibrahim SA. Antibacterial activity of red bell pepper against *Escherichia coli* O157: H7 in ground beef. Internet J Food Saf 2012; 14: 44-7.

[7]   Arora D, Sharma C, Jaglan S, *et al.* Advances in Endophytic Fungal Research. Cham, Switzerland: Springer International Publishing 2019.

[8]   Das A, Satyaprakash K. Antimicrobial properties of natural products: A review. Pharma Innov J 2018; 7: 532-7.

[9]   Ngwoke KG, Odimegwu DC, Esimone C. Antimicrobial natural products Science against microbial pathogens: communicating current research and technology advances Badajoz. Spain: FORMATEX 2011; p. 1011.

[10]   Hayek SAGR, Ibrahim SA. Antimicrobial Natural Products. In: Tew KD, Fisher PB, Eds. Advances in Cancer Research. San Diego, CA, Estados Unidos de América: Elsevier 2013; pp. 1–59.

[11]   Bobbarala V. Antimicrobial agents: BoD–Books on Demand. 2012.

[12]   Foster BC, Arnason JT, Briggs CJ. Natural health products and drug disposition. Annu Rev Pharmacol Toxicol 2005; 45(1): 203-26.
[http://dx.doi.org/10.1146/annurev.pharmtox.45.120403.095950] [PMID: 15822175]

[13]   Ayala-Zavala JF, Vega-Vega V, Rosas-Domínguez C, *et al.* Agro-industrial potential of exotic fruit byproducts as a source of food additives. Food Res Int 2011; 44(7): 1866-74.
[http://dx.doi.org/10.1016/j.foodres.2011.02.021]

[14]   Hartmann T. The lost origin of chemical ecology in the late 19th century. Proc Natl Acad Sci 2008; 105(12): 4541-6.
[http://dx.doi.org/10.1073/pnas.0709231105] [PMID: 18218780]

[15]   Tiwari BK, Valdramidis VP, O' Donnell CP, Muthukumarappan K, Bourke P, Cullen PJ. Application of natural antimicrobials for food preservation. J Agric Food Chem 2009; 57(14): 5987-6000.
[http://dx.doi.org/10.1021/jf900668n] [PMID: 19548681]

[16]   VanEtten HD, Mansfield JW, Bailey JA, Farmer EE. Two classes of plant antibiotics: Phytoalexins versus" phytoanticipins. Plant Cell 1994; 6(9): 1191-2.
[http://dx.doi.org/10.2307/3869817] [PMID: 12244269]

[17]   Franco CM, Vázquez BI. Natural compounds as antimicrobial agents. Antibiotics 2020; 9(5): 217.
[http://dx.doi.org/10.3390/antibiotics9050217] [PMID: 32365458]

[18]   Srinivasan D, Nathan S, Suresh T, Lakshmana Perumalsamy P. Antimicrobial activity of certain Indian medicinal plants used in folkloric medicine. J Ethnopharmacol 2001; 74(3): 217-20.
[http://dx.doi.org/10.1016/S0378-8741(00)00345-7] [PMID: 11274820]

[19]   Stefanović O. Antibacterial and antifungal activity of secondary metabolites of Teucrium species. Teucrium Species Biol Appl 2020; pp. 319-54.
[http://dx.doi.org/10.1007/978-3-030-52159-2_12]

[20]   Kwon DY, Choi JG, Kang OH, *et al.* *In vitro* and *in vivo* antibacterial activity of *punica granatum* peel ethanol extract against salmonella. Evidence-based Complement Altern Med 2011; 2011: 1-8.

[21]   Serda M, Becker FG, Cleary M, *et al.* Biotechnological strategies to improve safety and quality in food products. Uniw śląski 2017; 7: 343–354.

[22]   Cowan MM. Plant products as antimicrobial agents. Clin Microbiol Rev 1999; 12(4): 564-82.

[http://dx.doi.org/10.1128/CMR.12.4.564] [PMID: 10515903]

[23]   Murad T, Sabbagh G, Al-Kayali R. *In vitro* interaction of ciprofloxacin and some natural compounds against methicillin-resistant *staphylococcus aureus*. Innovare J Life Sci 2017; 5(1).

[24]   Saleem M, Durani AI, Asari A, *et al.* Investigation of antioxidant and antibacterial effects of citrus fruits peels extracts using different extracting agents: Phytochemical analysis with *in silico* studies. Heliyon 2023; 9(4): e15433.
[http://dx.doi.org/10.1016/j.heliyon.2023.e15433] [PMID: 37113773]

[25]   Dey D, Debnath S, Hazra S, Ghosh S, Ray R, Hazra B. Pomegranate pericarp extract enhances the antibacterial activity of ciprofloxacin against extended-spectrum β-lactamase (ESBL) and metallo-β-lactamase (MBL) producing Gram-negative bacilli. Food Chem Toxicol 2012; 50(12): 4302-9.
[http://dx.doi.org/10.1016/j.fct.2012.09.001] [PMID: 22982804]

[26]   Monika , Sharma DS, Kumar DN. Use of plant derived antimicrobials as an alternative to antibiotics. J Pharmacogn Phytochem 2020; 9(2): 1524-32.
[http://dx.doi.org/10.22271/phyto.2020.v9.i2y.11069]

[27]   Rahman MM, Rahaman MS, Islam MR, *et al.* Multifunctional therapeutic potential of phytocomplexes and natural extracts for antimicrobial properties. Antibiotics 2021; 10(9): 1076.
[http://dx.doi.org/10.3390/antibiotics10091076] [PMID: 34572660]

[28]   Ojala T. Biological screening of plant coumarins. 2001.

[29]   Ramawat KG, Mérillon JM. Natural products: Phytochemistry, botany and metabolism of alkaloids, phenolics and terpenes. Nat Prod Phytochem Bot Metab Alkaloids, Phenolics Terpenes 2013; pp. 1-4242.
[http://dx.doi.org/10.1007/978-3-642-22144-6]

[30]   Masyita A, Mustika Sari R, Dwi Astuti A, *et al.* Terpenes and terpenoids as main bioactive compounds of essential oils, their roles in human health and potential application as natural food preservatives. Food Chem X 2022; 13: 100217.
[http://dx.doi.org/10.1016/j.fochx.2022.100217] [PMID: 35498985]

[31]   Shan B, Cai YZ, Brooks JD, Corke H. The *in vitro* antibacterial activity of dietary spice and medicinal herb extracts. Int J Food Microbiol 2007; 117(1): 112-9.
[http://dx.doi.org/10.1016/j.ijfoodmicro.2007.03.003] [PMID: 17449125]

[32]   Hafidh RR, Abdulamir AS, Vern LS, *et al.* Inhibition of growth of highly resistant bacterial and fungal pathogens by a natural product. Open Microbiol J 2011; 5(1): 96-106.
[http://dx.doi.org/10.2174/1874285801105010096] [PMID: 21915230]

[33]   Oluwole DO, Coleman L, Buchanan W, Chen T, La Ragione RM, Liu LX. Antibiotics free compounds for chronic wound healing. Pharmaceutics 2022; 14(5): 1021.
[http://dx.doi.org/10.3390/pharmaceutics14051021] [PMID: 35631606]

[34]   Gutierrez J, Barry-Ryan C, Bourke P. The antimicrobial efficacy of plant essential oil combinations and interactions with food ingredients. Int J Food Microbiol 2008; 124(1): 91-7.
[http://dx.doi.org/10.1016/j.ijfoodmicro.2008.02.028] [PMID: 18378032]

[35]   Taylor M. Handbook of natural antimicrobials for food safety and quality. Elsevier 2014.

[36]   Arshad MS, Batool SA. Natural antimicrobials, their sources and food safety. Food additives 2017; 87(1).
[http://dx.doi.org/10.5772/intechopen.70197]

[37]   Du W-X, Avena-Bustillos RJ, Hua SST, McHugh TH. Antimicrobial volatile essential oils in edible films for food safety. Science against microbial pathogens: communicating current research and technological advances 2011; 2: 1124-34.

[38]   Raybaudi-Massilia RM, Mosqueda-Melgar J, Martín-Belloso O. Antimicrobial activity of essential oils on *Salmonella enteritidis*, *Escherichia coli*, and *Listeria innocua* in fruit juices. J Food Prot 2006;

69(7): 1579-86.
[http://dx.doi.org/10.4315/0362-028X-69.7.1579] [PMID: 16865889]

[39]   Turgis M, Han J, Caillet S, Lacroix M. Antimicrobial activity of mustard essential oil against *Escherichia coli* O157:H7 and *Salmonella typhi*. Food Control 2009; 20(12): 1073-9.
[http://dx.doi.org/10.1016/j.foodcont.2009.02.001]

[40]   Turgis M, Borsa J, Millette M, Salmieri S, Lacroix M. Effect of selected plant essential oils or their constituents and modified atmosphere packaging on the radiosensitivity of *Escherichia coli* O157:H7 and *Salmonella typhi* in ground beef. J Food Prot 2008; 71(3): 516-21.
[http://dx.doi.org/10.4315/0362-028X-71.3.516] [PMID: 18389694]

[41]   Pisoschi AM, Pop A, Georgescu C, Turcuş V, Olah NK, Mathe E. An overview of natural antimicrobials role in food. Eur J Med Chem 2018; 143: 922-35.
[http://dx.doi.org/10.1016/j.ejmech.2017.11.095] [PMID: 29227932]

[42]   Nanasombat S, Lohasupthawee P. Antibacterial activity of crude ethanolic extracts and essential oils of spices against Salmonellae and other enterobacteria. Curr Appl Sci Technol 2005; 5(3): 527-38.

[43]   Irkin R, Korukluoglu M. Growth inhibition of pathogenic bacteria and some yeasts by selected essential oils and survival of L. monocytogenes and C. albicans in apple-carrot juice. Foodborne Pathog Dis 2009; 6(3): 387-94.
[http://dx.doi.org/10.1089/fpd.2008.0195] [PMID: 19278342]

[44]   Burt SA, Reinders RD. Antibacterial activity of selected plant essential oils against *Escherichia coli* O157:H7. Lett Appl Microbiol 2003; 36(3): 162-7.
[http://dx.doi.org/10.1046/j.1472-765X.2003.01285.x] [PMID: 12581376]

[45]   Dorantes L, Colmenero R, Hernandez H, *et al.* Inhibition of growth of some foodborne pathogenic bacteria by Capsicum annum extracts. Int J Food Microbiol 2000; 57(1-2): 125-8.
[http://dx.doi.org/10.1016/S0168-1605(00)00216-6]

[46]   Teshome E, Forsido SF, Rupasinghe HPV, Olika Keyata E. Potentials of natural preservatives to enhance food safety and shelf life: A review. Sci Wor J 2022; 2022: 1-11.
[http://dx.doi.org/10.1155/2022/9901018] [PMID: 36193042]

[47]   Mandalari G, Bennett RN, Bisignano G, *et al.* Antimicrobial activity of flavonoids extracted from bergamot (*Citrus bergamia* Risso) peel, a byproduct of the essential oil industry. J Appl Microbiol 2007; 103(6): 2056-64.
[http://dx.doi.org/10.1111/j.1365-2672.2007.03456.x] [PMID: 18045389]

[48]   Park MJ, Choi WS, Kang HY, *et al.* Inhibitory effect of the essential oil from *Chamaecyparis obtusa* on the growth of food-borne pathogens. J Microbiol 2010; 48(4): 496-501.
[http://dx.doi.org/10.1007/s12275-010-9327-2] [PMID: 20799092]

[49]   McElhatton A, Sobral PJ do A. Novel technologies in food science: Their impact on products, consumer trends and the environment. Nov Technol Food Sci Their Impact Prod Consum Trends Environ 2012; pp. 1-421.
[http://dx.doi.org/10.1007/978-1-4419-7880-6]

[50]   Lee YL, Cesario T, Wang Y, Shanbrom E, Thrupp L. Antibacterial activity of vegetables and juices. Nutrition 2003; 19(11-12): 994-6.
[http://dx.doi.org/10.1016/j.nut.2003.08.003] [PMID: 14624951]

[51]   Davidson PM, Critzer FJ, Taylor TM. Naturally occurring antimicrobials for minimally processed foods. Annu Rev Food Sci Technol 2013; 4(1): 163-90.
[http://dx.doi.org/10.1146/annurev-food-030212-182535] [PMID: 23244398]

[52]   Ibrahim SA, Bor T, Song D, Tajkarimi M. Survival and growth characteristics of *Escherichia coli* O157:H7 in pomegranate-carrot and pomegranate-apple blend juices. Food Nutr Sci 2011; 2(8): 844-51.
[http://dx.doi.org/10.4236/fns.2011.28116]

[53] Al-Zoreky NS. Antimicrobial activity of pomegranate (*Punica granatum* L.) fruit peels. Int J Food Microbiol 2009; 134(3): 244-8.
[http://dx.doi.org/10.1016/j.ijfoodmicro.2009.07.002] [PMID: 19632734]

[54] Tajkarimi M, Ibrahim SA. Antimicrobial activity of ascorbic acid alone or in combination with lactic acid on *Escherichia coli* O157:H7 in laboratory medium and carrot juice. Food Control 2011; 22(6): 801-4.
[http://dx.doi.org/10.1016/j.foodcont.2010.11.030]

[55] Ibrahim SA, Yang G, Song D, Tse TSF. Antimicrobial effect of guava on *Escherichia coli* O157: H7 and *Salmonella typhimurium* in liquid medium. Int J Food Prop 2011; 14(1): 102-9.
[http://dx.doi.org/10.1080/10942910903147833]

[56] Hayek SA, Ibrahim SA. Ibrahim, Antimicrobial activity of xoconostle pears (*Opuntia matudae*) against *Escherichia coli* O H7 in laboratory medium. Int J Microbiol 2012; 2012: 1-6.
[http://dx.doi.org/10.1155/2012/368472] [PMID: 22934117]

[57] Fattouch S, Caboni P, Coroneo V, *et al.* Antimicrobial activity of *Tunisian quince* (*Cydonia oblonga* Miller) pulp and peel polyphenolic extracts. J Agric Food Chem 2007; 55(3): 963-9.
[http://dx.doi.org/10.1021/jf062614e] [PMID: 17263500]

[58] Belguith H, Kthiri F, Ben Ammar A, Jaafoura H, Ben Hamida J, Landoulsi A. Morphological and biochemical changes of Salmonella hadar exposed to aqueous garlic extract. Int J Morphol 2009; 27(3): 705-13.
[http://dx.doi.org/10.4067/S0717-95022009000300013]

[59] Avato P, Tursi F, Vitali C, Miccolis V, Candido V. Allylsulfide constituents of garlic volatile oil as antimicrobial agents. Phytomedicine 2000; 7(3): 239-43.
[http://dx.doi.org/10.1016/S0944-7113(00)80010-0] [PMID: 11185736]

[60] Dorman HJD, Deans SG. Antimicrobial agents from plants: Antibacterial activity of plant volatile oils. J Appl Microbiol 2000; 88(2): 308-16.
[http://dx.doi.org/10.1046/j.1365-2672.2000.00969.x] [PMID: 10736000]

[61] Pei J, Mei J, Yu H, Qiu W, Xie J. effect of gum tragacanth-sodium alginate active coatings incorporated with epigallocatechin gallate and lysozyme on the quality of large yellow croaker at superchilling condition. Front Nutr 2022; 8: 812741.
[http://dx.doi.org/10.3389/fnut.2021.812741] [PMID: 35118111]

[62] Dubey A, Mishra N, Singh N. Antimicrobial activity of some selected vegetables. Int J Appl Biol Pharm Technol 2010; 1(3): 994-9.

[63] Delaquis P, Stanich K, Girard B, Mazza G. Antimicrobial activity of individual and mixed fractions of dill, cilantro, coriander and eucalyptus essential oils. Int J Food Microbiol 2002; 74(1-2): 101-9.
[http://dx.doi.org/10.1016/S0168-1605(01)00734-6] [PMID: 11929164]

[64] Zorić N, Kosalec I. The Antimicrobial Activities of Oleuropein and Hydroxytyrosol. Promis Antimicrob from Nat Prod 2022; pp. 75-89.

[65] Markin D, Duek L, Berdicevsky I. *In vitro* antimicrobial activity of olive leaves. Antimikrobielle Wirksamkeit von Olivenblättern *in vitro*. Mycoses 2003; 46(3-4): 132-6.
[http://dx.doi.org/10.1046/j.1439-0507.2003.00859.x] [PMID: 12870202]

[66] Pereira A, Ferreira I, Marcelino F, *et al.* Phenolic compounds and antimicrobial activity of olive (*Olea europaea* L. Cv. Cobrançosa) leaves. Molecules 2007; 12(5): 1153-62.
[http://dx.doi.org/10.3390/12051153] [PMID: 17873849]

[67] Khayat S, Al-Zahrani SHM, Basudan N, *et al.* Chemical composition and *in vitro* antibacterial activities of traditional medicinal plant: Olea sp. Biomed Res 2018; 29: 1037-47.

[68] Singh J, Meshram V, Gupta M. Bioactive natural products in drug discovery. Bioact Nat Prod Drug Discov 2020; pp. 1-733.

[http://dx.doi.org/10.1007/978-981-15-1394-7]

[69]   Thiem B, Goślińska O. Antimicrobial activity of *Rubus chamaemorus* leaves. Fitoterapia 2004; 75(1): 93-5.
         [http://dx.doi.org/10.1016/j.fitote.2003.08.014] [PMID: 14693229]

[70]   Dogasaki C, Shindo T, Furuhata K, Fukuyama M. Identification of chemical structure of antibacterial components against *Legionella pneumophila* in a coffee beverage. Yakugaku Zasshi 2002; 122(7): 487-94.
         [http://dx.doi.org/10.1248/yakushi.122.487] [PMID: 12136645]

[71]   Ibrahim S, Salameh M, Phetsomphou S, Yang H, Seo C. Application of caffeine, 1,3,7-trimethylxanthine, to control *Escherichia coli* O157:H7. Food Chem 2006; 99(4): 645-50.
         [http://dx.doi.org/10.1016/j.foodchem.2005.08.026]

[72]   Duda-Chodak A, Tarko T, Petka-Poniatowska K. Antimicrobial compounds in food packaging. J Mol Sci 2023; 24: 2457.
         [http://dx.doi.org/10.3390/ijms24032457]

[73]   Kim S, Fung DYC. Antibacterial effect of crude water-soluble arrowroot (*Puerariae radix*) tea extracts on food-borne pathogens in liquid medium. Lett Appl Microbiol 2004; 39(4): 319-25.
         [http://dx.doi.org/10.1111/j.1472-765X.2004.01582.x] [PMID: 15355532]

[74]   Burt S. Essential oils: their antibacterial properties and potential applications in foods--a review. Int J Food Microbiol 2004; 94(3): 223-53.
         [http://dx.doi.org/[https://doi.org/10.1016/j.ijfoodmicro.2004.03.022]] [PMID: [PMID: 15246235]]

[75]   Shimamura T, Zhao WH, Hu ZQ. Mechanism of action and potential for use of tea catechin as an antiinfective agent. Antiinfect Agents Med Chem 2007; 6(1): 57-62.
         [http://dx.doi.org/10.2174/187152107779314124]

[76]   Hoskin DW, Ramamoorthy A. Studies on anticancer activities of antimicrobial peptides. Biochim Biophys Acta Biomembr 2008; 1778(2): 357-75.
         [http://dx.doi.org/10.1016/j.bbamem.2007.11.008] [PMID: 18078805]

[77]   Jain V, Benyoucef L, Jain V, *et al.* New Approaches for Modeling and Evaluating Agility in Integrated Supply Chains.Supply Chain. Rijeka: IntechOpen 2009.

[78]   Dell'olmo E. Host Defence Peptides (HDPs) as novel biopreservatives in food industry applications. Università degli Studi di Napoli Federico II 2020.

[79]   Burrowes OJ, Hadjicharalambous C, Diamond G, Lee T-C. Evaluation of antimicrobial spectrum and cytotoxic activity of pleurocidin for food applications. J Food Sci 2004; 69(3): FMS66-71.
         [http://dx.doi.org/10.1111/j.1365-2621.2004.tb13373.x]

[80]   Cole AM, Darouiche RO, Legarda D, Connell N, Diamond G. Characterization of a fish antimicrobial peptide: gene expression, subcellular localization, and spectrum of activity. Antimicrob Agents Chemother 2000; 44(8): 2039-45.
         [http://dx.doi.org/10.1128/AAC.44.8.2039-2045.2000] [PMID: 10898673]

[81]   Potter R, Truelstruphansen L, Gill T. Inhibition of foodborne bacteria by native and modified protamine: Importance of electrostatic interactions. Int J Food Microbiol 2005; 103(1): 23-34.
         [http://dx.doi.org/10.1016/j.ijfoodmicro.2004.12.019] [PMID: 16084263]

[82]   Humblot V, Yala JF, Thebault P, *et al.* The antibacterial activity of Magainin I immobilized onto mixed thiols Self-Assembled Monolayers. Biomaterials 2009; 30(21): 3503-12.
         [http://dx.doi.org/10.1016/j.biomaterials.2009.03.025] [PMID: 19345992]

[83]   Seifu E, Buys EM, Donkin EF. Significance of the lactoperoxidase system in the dairy industry and its potential applications: A review. Trends Food Sci Technol 2005; 16(4): 137-54.
         [http://dx.doi.org/10.1016/j.tifs.2004.11.002]

[84]   Lönnerdal B. Nutritional and physiologic significance of human milk proteins. Am J Clin Nutr 2003;

77(6): 1537S-43S.
[http://dx.doi.org/10.1093/ajcn/77.6.1537S] [PMID: 12812151]

[85]   Lee NK, Paik HD. Status, antimicrobial mechanism, and regulation of natural preservatives in livestock food systems. Han-gug Chugsan Sigpum Hag-hoeji 2016; 36(4): 547-57.
[http://dx.doi.org/10.5851/kosfa.2016.36.4.547] [PMID: 27621697]

[86]   Ben-Shalom N, Ardi R, Pinto R, Aki C, Fallik E. Controlling gray mould caused by Botrytis cinerea in cucumber plants by means of chitosan. Crop Prot 2003; 22(2): 285-90.
[http://dx.doi.org/10.1016/S0261-2194(02)00149-7]

[87]   Hasan S, Boddu VM, Viswanath DS, *et al.* Application of chitosan in textiles. 2022; 323-37.
[http://dx.doi.org/10.1007/978-3-031-01229-7_9]

[88]   Fernandes JC, Tavaria FK, Soares JC, *et al.* Antimicrobial effects of chitosans and chitooligosaccharides, upon *Staphylococcus aureus* and *Escherichia coli*, in food model systems. Food Microbiol 2008; 25(7): 922-8.
[http://dx.doi.org/10.1016/j.fm.2008.05.003] [PMID: 18721683]

[89]   Demain AL. Pharmaceutically active secondary metabolites of microorganisms. Appl Microbiol Biotechnol 1999; 52(4): 455-63.
[http://dx.doi.org/10.1007/s002530051546] [PMID: 10570792]

[90]   Nes IF, Johnsborg O. Exploration of antimicrobial potential in LAB by genomics. Curr Opin Biotechnol 2004; 15(2): 100-4.
[http://dx.doi.org/10.1016/j.copbio.2004.02.001] [PMID: 15081046]

[91]   HUISUO H. Applications of lactic acid and its derivatives in meat products and methods to analyze related additives in restructured meat. University of Missouri-Columbia 2016.

[92]   Skrivanova E, Marounek M, Benda V, Brezina P. Susceptibility of *Escherichia coli*, *Salmonella* sp. and *Clostridium perfringensto* organic acids and monolaurin. Vet Med 2006; 51(3): 81-8.
[http://dx.doi.org/10.17221/5524-VETMED]

[93]   Raftari M. Antibacterial activity of organic acids on the growth of selected bacteria in meat samples. Lambert Academic Publishing 2009.

[94]   Maherani B, Ayari S, Lacroix M. The use of natural antimicrobials combined with nonthermal treatments to control human pathogens. ACS Symp Ser 2018; 1287: 149-69.
[http://dx.doi.org/10.1021/bk-2018-1287.ch008]

[95]   Muthukumarasamy P, Holley R. Survival of *Escherichia coli* O157:H7 in dry fermented sausages containing micro-encapsulated probiotic lactic acid bacteria. Food Microbiol 2007; 24(1): 82-8.
[http://dx.doi.org/10.1016/j.fm.2006.03.004] [PMID: 16943098]

[96]   Gyawali R, Ibrahim SA, Abu Hasfa SH, Smqadri SQ, Haik Y. Antimicrobial Activity of Copper Alone and in Combination with Lactic Acid against *Escherichia coli* O157:H7 in Laboratory Medium and on the Surface of Lettuce and Tomatoes. J Pathogens 2011; 2011: 1-9.
[http://dx.doi.org/10.4061/2011/650968] [PMID: 22567336]

[97]   Choi O, Hu Z. Size dependent and reactive oxygen species related nanosilver toxicity to nitrifying bacteria. Environ Sci Technol 2008; 42(12): 4583-8.
[http://dx.doi.org/10.1021/es703238h] [PMID: 18605590]

[98]   Turovskiy Y, Sutyak Noll K, Chikindas ML. The aetiology of bacterial vaginosis. J Appl Microbiol 2011; 110(5): 1105-28.
[http://dx.doi.org/10.1111/j.1365-2672.2011.04977.x] [PMID: 21332897]

[99]   Reid G. Probiotic Lactobacilli for urogenital health in women. J Clin Gastroenterol 2008; 42(3) (3): S234-6.
[http://dx.doi.org/10.1097/MCG.0b013e31817f1298] [PMID: 18685506]

[100] Šušković J, Kos B, Beganović J, *et al.* Antimicrobial activity : The most important property of

probiotic and starter lactic acid bacteria. Food Technol Biotechnol 2010; 48(3): 296-307.

[101] Savadogo A, Ouattara CAT, Bassole IHN, *et al.* Bacteriocins and lactic acid bacteria : A mini review. Afr J Biotechnol 2006; 5(9): 678-84.

[102] Hayek SA, Ibrahim SA. Antimicrobial activity of xoconostle pears (*Opuntia matudae*) against *Escherichia coli* O H7 in laboratory medium. Int J Microbiol 2012; 2012: 1-6.
[http://dx.doi.org/10.1155/2012/368472]

[103] Gálvez A, López RL, Abriouel H, Valdivia E, Omar NB. Application of bacteriocins in the control of foodborne pathogenic and spoilage bacteria. Crit Rev Biotechnol 2008; 28(2): 125-52.
[http://dx.doi.org/10.1080/07388550802107202] [PMID: 18568851]

[104] Singh T, Pandove G, Arora M. Bacteriocins JBS 2013; 1(2): 73-5.

[105] Available from: https://www.sciencedirect.com/science/article/pii/B9780128113721000026
[http://dx.doi.org/10.1016/B978-0-12-811372-1.00002-6]

[106] Hancock REW, Lehrer R. Cationic peptides: A new source of antibiotics. Trends Biotechnol 1998; 16(2): 82-8.
[http://dx.doi.org/10.1016/S0167-7799(97)01156-6] [PMID: 9487736]

[107] Papagianni M S. Anastasiadou, Pediocins: The bacteriocins of Pediococci. Microb Cell factories 2009; 8(1): 1-16.

[108] de Oliveira TM, de Fátima Ferreira Soares N, Pereira RM, de Freitas Fraga K. Development and evaluation of antimicrobial natamycin-incorporated film in gorgonzola cheese conservation. Packag Technol Sci 2007; 20(2): 147-53.
[http://dx.doi.org/10.1002/pts.756]

[109] Otunba by, Adebisi A. Probiotic Properties of Pediococcus Species Isolated from Sorghum. University of Lagos 2017.

[110] Doleyres Y, Beck P, Vollenweider S, Lacroix C. Production of 3-hydroxypropionaldehyde using a two step process with *Lactobacillus reuteri.* Appl Microbiol Biotechnol 2005; 68(4): 467-74.
[http://dx.doi.org/10.1007/s00253-005-1895-4] [PMID: 15682289]

[111] Gänzle MG, Vogel RF. Studies on the mode of action of reutericyclin. Appl Environ Microbiol 2003; 69(2): 1305-7.
[http://dx.doi.org/10.1128/AEM.69.2.1305-1307.2003] [PMID: 12571063]

[112] Axelsson L. Production of a broad spectrum antimicrobial substance by *Lactobacillus reuteri.* Microb Ecol Health Dis 2: 131-6.

[113] Arqués JL, Fernández J, Gaya P, Nuñez M, Rodríguez E, Medina M. Antimicrobial activity of reuterin in combination with nisin against food-borne pathogens. Int J Food Microbiol 2004; 95(2): 225-9.
[http://dx.doi.org/10.1016/j.ijfoodmicro.2004.03.009] [PMID: 15282134]

[114] Sakharkar P, Pati T. Antimicrobial activity of *Cassia alata.* Indian J Pharm Sci 1998; 60: 311-2.

[115] Rathore M, Sharma K, Sharma N. Antimicrobial potential of botanicals and disease control. Nat Prod J 2012; 1(2): 105-15.
[http://dx.doi.org/10.2174/2210316311101020105]

[116] Nithya V. Antimicrobial activity of medicinal plants. Inven Rapid Ethnopharmacol 2021; 2(1): 2-5.

[117] Nair A, Bhide S S. Antimicrobial properties of different parts of *Semecarpus anacardium.* Indian drugs 1996; 33(7): 323-8.

[118] Jain SK. Medicinal plants. 5th., New Delhi: National Book trust 2012.

[119] Pankajalakshmi Venugopal, Taralakshmi V, *et al.* Antidermatophytic activity of neem (*Azadirachta indica*) leaves *in vitro.* Indian J Pharmacol 1994; 26(2): 141-3.

[120] Rai MK. *In vitro* evaluation of medicinal plant extracts against *Pestalotiopsis mangiferae.* Hindustan

Antibiot 1996; 38(1-4): 53-6.

[121] Rana BK, Singh UP, Taneja V. Antifungal activity and kinetics of inhibition by essential oil isolated from leaves of *Aegle marmelos*. J Ethnopharmacol 1997; 57(1): 29-34.
[http://dx.doi.org/10.1016/S0378-8741(97)00044-5] [PMID: 9234162]

[122] Sakharkar PR, Patil AT. Antifungal activity of *Cassia alata*. Hamdard Med 1998; 41(3): 20-1.

[123] Dhuley JN. Therapeutic efficacy of Ashwagandha against experimental aspergillosis in mice. Immunopharmacol Immunotoxicol 1998; 20(1): 191-8.
[http://dx.doi.org/10.3109/08923979809034817] [PMID: 9543708]

[124] Lalita B. In vitro studies on the effect of glycyrrhizin from Indian *Glycyrrhiza glabra* Linn on some RNA and DNA viruses. Indian J Pharmacol 1994; 26(3): 194-9.

[125] Premanathan M, Nakashima H, Kathiresan K, Rajendran N, Yamamoto N. *In vitro* anti human immunodeficiency virus activity of mangrove plants. Indian J Med Res 1996; 103: 278-81.
[PMID: 8707365]

[126] Chatterjee S, Das S. Anti-arthritic and anti-inflammatory effect of a poly-herbal drug (EASE)@: Its mechanism of action. Indian J Pharmacol 1996; 28(2): 116-9.

[127] Valecha N, Biswas S, Badoni V, *et al*. Antimalarial activity of Artemisia japonica, *Artemisia maritima* and *Artemisia nilegarica*. Indian J Pharmacol 1994; 26(2): 144-6.

[128] Ray S, Majumder HK, Chakravarty AK, Mukhopadhyay S, Gil RR, Cordell GA. Amarogentin, a naturally occurring secoiridoid glycoside and a newly recognized inhibitor of topoisomerase I from *Leishmania donovani*. J Nat Prod 1996; 59(1): 27-9.
[http://dx.doi.org/10.1021/np960018g] [PMID: 8984149]

[129] Talakal TS, Dwivedi SK, Sharma SR. *In vitro* and *in vivo* therapeutic activity of *Parthenium hysterophorus* against *Trypanosoma evansi*. Indian J Exp Biol 1995; 33(11): 894-6.
[PMID: 8786167]

[130] Ciocan D, Ioan B. Plant products as antimicrobial agents. Analele Stiint ale Univ Alexandru Ioan Cuza din Iasi Sec II a Genet Biol Mol; 8.

[131] Çelenk BM, Handan AA, Tokatli E, *et al*. Resistance properties and control of *Alicyclobacillus acidoterrestris*. İzmir Institute of Technology 2014.

[132] Bajpai VK, Rahman A, Dung NT, Huh MK, Kang SC. *In vitro* inhibition of food spoilage and foodborne pathogenic bacteria by essential oil and leaf extracts of *Magnolia liliflora* Desr. J Food Sci 2008; 73(6): M314-20.
[http://dx.doi.org/10.1111/j.1750-3841.2008.00841.x] [PMID: 19241564]

[133] Tafesh A, Najami N, Jadoun J, Halahlih F, Riepl H, Azaizeh H. Synergistic antibacterial effects of polyphenolic compounds from olive mill wastewater. Evid Based Complement Alternat Med 2011; 2011: 1-9.
[http://dx.doi.org/10.1155/2011/431021] [PMID: 21647315]

[134] Brogden KA. Antimicrobial peptides: pore formers or metabolic inhibitors in bacteria? Nat Rev Microbiol 2005; 3(3): 238-50.
[http://dx.doi.org/10.1038/nrmicro1098] [PMID: 15703760]

[135] Valladão ABG, Torres AG, Freire DMG, Cammarota MC. Profiles of fatty acids and triacylglycerols and their influence on the anaerobic biodegradability of effluents from poultry slaughterhouse. Bioresour Technol 2011; 102(14): 7043-50.
[http://dx.doi.org/10.1016/j.biortech.2011.04.037] [PMID: 21576016]

[136] Hayek S. Use of sweet potato to develop a medium for cultivation of lactic acid bacteria. North Carolina Agricultural and Technical State University 2013.

[137] Desbois AP, Smith VJ. Antibacterial free fatty acids: Activities, mechanisms of action and biotechnological potential. Appl Microbiol Biotechnol 2010; 85(6): 1629-42.

[http://dx.doi.org/10.1007/s00253-009-2355-3] [PMID: 19956944]

[138] Kankaanpää PE, Salminen SJ, Isolauri E, Lee YK. The influence of polyunsaturated fatty acids on probiotic growth and adhesion. FEMS Microbiol Lett 2001; 194(2): 149-53.
[http://dx.doi.org/10.1111/j.1574-6968.2001.tb09460.x] [PMID: 11164299]

[139] Driessen AJM, van den Hooven HW, Kuiper W, *et al.* Mechanistic studies of lantibiotic-induced permeabilization of phospholipid vesicles. Biochemistry 1995; 34(5): 1606-14.
[http://dx.doi.org/10.1021/bi00005a017] [PMID: 7849020]

[140] Lee DU, Heinz V, Knorr D. Effects of combination treatments of nisin and high-intensity ultrasound with high pressure on the microbial inactivation in liquid whole egg. Innov Food Sci Emerg Technol 2003; 4(4): 387-93.
[http://dx.doi.org/10.1016/S1466-8564(03)00039-0]

[141] Crandall AD, Montville TJ. Nisin resistance in Listeria monocytogenes ATCC 700302 is a complex phenotype. Appl Environ Microbiol 1998; 64(1): 231-7.
[http://dx.doi.org/10.1128/AEM.64.1.231-237.1998] [PMID: 9435079]

[142] De Vuyst L, Leroy F. Bacteriocins from lactic acid bacteria: production, purification, and food applications. J Mol Microbiol Biotechnol 2007; 13(4): 194-9.
[PMID: 17827969]

[143] Soares MO, Vinha AF, Coutinho F, *et al.* Antimicrobial natural products.Microbial pathogens and strategies for combating them: science, technology and education. FORMATEX 2013; pp. 946-50.

[144] Burt S. Essential oils: Their antibacterial properties and potential applications in foods : A review. Int J Food Microbiol 2004; 94(3): 223-53.
[http://dx.doi.org/10.1016/j.ijfoodmicro.2004.03.022] [PMID: 15246235]

[145] Lin YC, Chou CC. Effect of heat treatment on total phenolic and anthocyanin contents as well as antioxidant activity of the extract from *Aspergillus awamori*- fermented black soybeans, a healthy food ingredient. Int J Food Sci Nutr 2009; 60(7): 627-36.
[http://dx.doi.org/10.3109/09637480801992492] [PMID: 19817642]

[146] Kim TJ, Silva JL, Jung YS. Antibacterial activity of fresh and processed red muscadine juice and the role of their polar compounds on *Escherichia coli* O157:H7. J Appl Microbiol 2009; 107(2): 533-9.
[http://dx.doi.org/10.1111/j.1365-2672.2009.04239.x] [PMID: 19302492]

[147] Kim TJ, Weng WL, Stojanovic J, Lu Y, Jung YS, Silva JL. Antimicrobial effect of water-soluble muscadine seed extracts on *Escherichia coli* O157:H7. J Food Prot 2008; 71(7): 1465-8.
[http://dx.doi.org/10.4315/0362-028X-71.7.1465] [PMID: 18680948]

[148] Niino H, Sakane I, Okanoya K, Kuribayashi S, Kinugasa H. Determination of mechanism of flock sediment formation in tea beverages. J Agric Food Chem 2005; 53(10): 3995-9.
[http://dx.doi.org/10.1021/jf047904e] [PMID: 15884829]

[149] Álvarez-Martínez FJ, Barrajón-Catalán E, Micol V. Tackling antibiotic resistance with compounds of natural origin: A comprehensive review. Biomedicines 2020; 8(10): 405.
[http://dx.doi.org/10.3390/biomedicines8100405] [PMID: 33050619]

[150] Gupta PD, Birdi TJ. Development of botanicals to combat antibiotic resistance. J Ayurveda Integr Med 2017; 8(4): 266-75.
[http://dx.doi.org/10.1016/j.jaim.2017.05.004] [PMID: 28869082]

[151] San Millan A, MacLean RC. Fitness costs of plasmids a limit to plasmid transmission. Microbiol Spectr 2017; 5(5): 5.5.02.
[http://dx.doi.org/10.1128/microbiolspec.MTBP-0016-2017] [PMID: 28944751]

[152] Durão P, Balbontín R, Gordo I. Evolutionary mechanisms shaping the maintenance of antibiotic resistance. Trends Microbiol 2018; 26(8): 677-91.
[http://dx.doi.org/10.1016/j.tim.2018.01.005] [PMID: 29439838]

[153] Melnyk AH, Wong A, Kassen R. The fitness costs of antibiotic resistance mutations. Evol Appl 2015; 8(3): 273-83.
[http://dx.doi.org/10.1111/eva.12196] [PMID: 25861385]

[154] Pancu DF, Scurtu A, Macasoi IG, *et al.* Antibiotics: Conventional therapy and natural compounds with antibacterial activity a pharmaco-toxicological screening. Antibiotics 2021; 10(4): 401.
[http://dx.doi.org/10.3390/antibiotics10040401] [PMID: 33917092]

[155] Sharifi-Rad J. Herbal Antibiotics: Moving back into the mainstream as an alternative for "Superbugs". Cell Mol Biol 2016; 62(9): 1-2.
[PMID: 27585254]

[156] Everts S. Antibiotic side effects explained. Chem Eng News 2013; 91(21): 9.
[http://dx.doi.org/10.1021/cen-09121-notw4]

CHAPTER 8

Human Diseases and Recent Biotechnology Breakthroughs in Curbing Diseases

Ana K. Villagómez-Guzmán[1,*] and **Israel Valencia Quiroz**[2]

[1] *Laboratory of Natural Products Bioactivity, UBIPRO, Superior Studies Faculty (FES)-Iztacala, National Autonomous University of Mexico (UNAM), Tlalnepantla de Baz, México State, 54090, México*

[2] *Phytochemistry Laboratory, UBIPRO, Superior Studies Faculty (FES)-Iztacala, National Autonomous University of Mexico (UNAM), Tlalnepantla de Baz, México State, 54090, México*

Abstract: Medical biotechnology represents a field in continuous progress and today has revolutionized how illnesses are diagnosed and treated. A look at the latest medical biotechnological breakthroughs shows how biotechnology innovations are changing medicine. Recently, we saw how biotechnology affected efforts to combat the coronavirus disease 2019 (COVID-19) pandemic on the world's health. The scientific community has been working assiduously to develop effective treatments for the prevention and management of other diseases, such as cancer, human immunodeficiency virus (HIV), cardiovascular disease, diabetes mellitus, and neurodegenerative disorders such as Alzheimer's disease, along with other dementia variants that stand out among the leading causes of mortality worldwide. This effort has recently resulted in the development of RNA vaccines. Some of the most promising biotechnological developments include gene therapy to alter an individual's genetic makeup through diverse techniques, immunotherapeutic methods that bolster the body's natural immune defense mechanisms, and precision medicine strategies in which treatment is personalized to a patient's genetic profile. This chapter provides an overview of the most prevalent and deadly human diseases with a focus on recent biotechnological breakthroughs.

Keywords: Biotechnological breakthroughs, Gene therapy, Immunotherapy, Medical biotechnology, Precision medicine.

INTRODUCTION

In recent years, medical biotechnology has emerged as a highly progressive discipline that utilizes living organisms, cells, cell products, or materials to gene-

---

* **Corresponding author Ana K. Villagómez-Guzmán:** Laboratory of Natural Products Bioactivity, UBIPRO, Superior Studies Faculty (FES)-Iztacala, National Autonomous University of Mexico (UNAM), Tlalnepantla de Baz, México State, 54090, México; Tel: +525572731888; E-mail: kvillagomez@live.com

Israel Valencia Quiroz (Ed.)

rate innovative healthcare solutions [1]. Biotechnology enterprises and the scientific community are employing the most recent advancements in molecular biology, genetics, and nanotechnology to fabricate innovative medicinal products, vaccines, diagnostic instruments, and other healthcare commodities to improve human welfare. The present role of medical biotechnology can be seen in precision medicine and the utilization of progressive technologies such as gene manipulation and cellular therapy (Fig. 1) [2]. Precision medicine can potentially revolutionize the healthcare industry for both collectives and individuals by enabling early disease detection, enhancing diagnostic accuracy, and customizing therapeutic interventions [3].

This field of study employs various essential technologies, such as big data, artificial intelligence, diverse omics, and pharmaco-omics, as well as environmental and social factors. Additionally, it involves integrating these technologies with tools from preventive and population medicine, all of which are critical components of this field [4]. A personalized approach to medicine needs to be supported by the utilization of cutting-edge technologies, including gene manipulation. The landscape of gene therapy technology has advanced significantly [5].

A gene-editing method called clustered regularly interspaced short palindromic repeats (CRISPR) was developed for increased robustness and is based on the innate immune system of bacteria. Due to its increased specificity and effectiveness, CRISPR/Cas9 has been widely used in the treatment of a variety of genetic and nongenetic diseases, including but not limited to cancer, genetic hemolytic diseases, acquired immunodeficiency syndrome, cardiovascular disease (CVD), ocular diseases, neurodegenerative diseases, and some X-linked diseases. Additionally, some researchers have used the CRISPR/Cas9 methodology in the context of gene therapy and immune therapy for cancer treatment to treat or mitigate cancer [6]. Cancer immune therapy is a biological procedure that uses several immune system elements to protect the host from the development of primary tumors or improve the chances of cancer evasion. Cancer immunotherapy has led to significant improvements for patients in terms of survival and quality of life when compared to chemotherapy, radiotherapy, and surgery. Immunotherapy has recently gained acceptance in the field of cancer treatment as a cutting-edge foundational therapy, applicable from the metastatic stage to the adjuvant and neoadjuvant environments in a variety of disease categories (Fig. 1) [7].

From developing new treatments for diseases to creating diagnostic tools that can detect illnesses earlier, medical biotechnology significantly impacts how we approach healthcare. Overall, the current state of biotechnology in healthcare is dynamic and ever-evolving, and the field continues to push the boundaries of medical science. The prospect of utilizing gene therapy to treat chronic ailments necessitating lifelong care and medical supervision has sparked scientific interest. CVD and neurodegenerative diseases are considered appropriate models for the comprehensive utilization of precision medicine technologies throughout the various stages of disease progression [8].

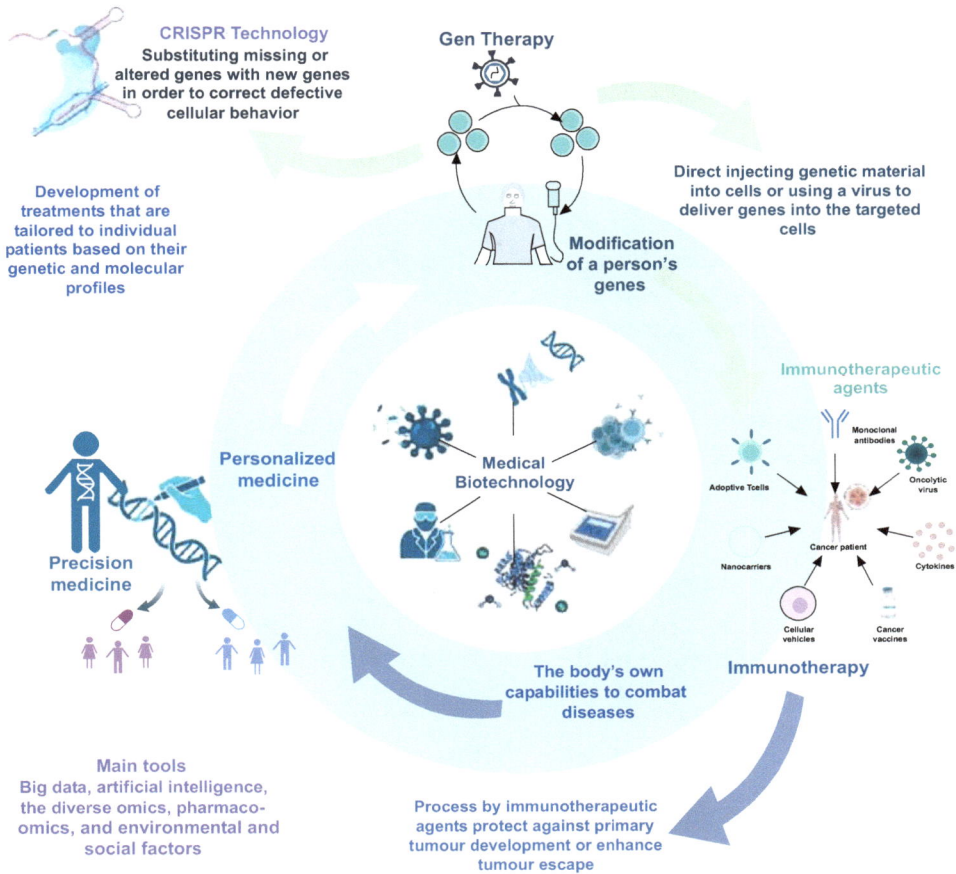

Fig. (1). Medical biotechnology as a central scientific area for the development of promising advancements in targeted therapies such as gene therapy, precision medicine and immunotherapies. Created with BioRender.com.

In 2019, noncommunicable diseases were responsible for 74.36% of all deaths worldwide. CVDs, neoplasms, and chronic respiratory disorders (CRDs) are a few of the prominent non-communicable diseases (NCDs). In 2019, deaths from these three diseases made up a sizable fraction of deaths worldwide from all causes [9].

Utilizing the strength of biological methods and processes, novel medicinal therapies are being created through biotechnology. As one of the most promising fields of recent research, biotechnology represents a profound intersection of science and medicine and has the potential to significantly change healthcare and our way of life.

CVD: THE LEADING CAUSE OF DEATH WORLDWIDE

CVD, a term encompassing a spectrum of disorders affecting the structural and functional attributes of the heart and blood vessels, is the leading cause of death worldwide. As the World Health Organization states, ischemic heart disease (IHD) alone accounted for nearly 7.4 million deaths globally in 2012, thereby establishing itself as the primary cause of mortality [10]. Symptoms of IHD, also known as coronary artery disease, involve reduced blood flow to the heart, and IHD is often precipitated by numerous factors. These include high cholesterol, hypertension, prolonged diabetes, abdominal obesity, inadequate physical activity, poor fruit and vegetable intake, excessive alcohol and tobacco use, stress, and more [10].

While the treatment for IHD typically encompasses medication, lifestyle alterations, and potentially surgical intervention, recent advancements in the field have led to novel therapeutic prospects. The development of recombinant nematode anticoagulant protein c2 (rNAPc2) therapy, in particular, has shown promise in preventing the initiation of the coagulation cascade, namely, the activated factor VIIa/tissue factor complex (FVIIa/TF) [11]. This is particularly significant as it hinders excessive blood clotting, a condition that could lead to severe medical issues such as deep vein thrombosis, pulmonary embolism, and stroke.

Over the last few decades, stem cell therapy has emerged as a revolutionary therapeutic strategy for managing heart conditions. Recent empirical investigations and clinical trials suggest that cell-centric therapies could potentially enhance cardiac functionality, and the prospects of cardiac regeneration from these treatments are garnering considerable interest [12]. Outcomes of stem cell therapy implementation are manifold, encompassing the reduction of scarring, promotion of angiogenesis, prevention or reduction of myocardial cell deterioration, and improvement of cardiac function [12].

Angiogenesis, a physiological phenomenon that triggers the formation of new blood vessels, can play a pivotal role in cardiac health. However, if this complex process is mismanaged, it could contribute to the onset of various pathological conditions, including cancer, ischemia disorders, inflammatory diseases, infectious diseases, and immunological dysfunctions [13]. Nevertheless, the pursuit of "therapeutic angiogenesis" as a solution for ischemic ailments has yet to yield consistent results, even with extensive preclinical and clinical efforts [14]. Nevertheless, the promise of these innovative therapies heralds a new era in the management and treatment of CVD.

CANCER: ONE OF THE MOST COMMON AND DEVASTATING DISEASES

The human body is composed of billions of specialized cells, forming the bedrock of our physical existence and functioning. Each of these cells has a life cycle essential for the maintenance of our biological systems. However, daily exposure to various external factors can disrupt this cellular life cycle, fostering anomalies such as unregulated cell division. This process manifests as the disease we recognize as cancer.

In 2008, a staggering 12 million new cases of cancer were reported, and it was determined that infectious agents were responsible for 16% of these occurrences, equating to nearly 2 million cases [15]. The proportion of cancer cases attributed to infectious agents, also known as the population-attributable fraction (PAF), varied by region. In developed countries, this figure was approximately 7.4%, while in less developed countries, it was substantially higher, at 22.9%. Australia and New Zealand demonstrated a lower PAF at 3.3%, while sub-Saharan Africa saw a significantly larger proportion at 32.7%.

Further analysis revealed that most of these infection-associated cancer cases were primarily due to three pathogens: *Helicobacter pylori*, hepatitis B and C viruses, and human papillomaviruses. These pathogens were chiefly implicated in gastric, hepatic, and cervical-uterine cancers, accounting for approximately 9 million cases.

Notably, sex disparity was observed in the distribution of infection-related cancers. While nearly half of such cancer cases in women were attributed to cervical and uterine cancers, over 80% of cases in men were hepatic and gastric cancers. Furthermore, the burden of infection-related cancers was especially high in younger demographics, with these conditions accounting for approximately 30% of cases in individuals under 50 years [15]. This underscores the pivotal role of infectious agents in the etiology of cancer and the importance of targeted prevention and treatment strategies.

DIABETES: ONE OF THE MOST PREVALENT AND DEADLY HUMAN DISEASES

The global incidence of type 2 diabetes represents a substantial public health crisis, significantly straining the economies of all nations, with developing countries being particularly affected. The quickening pace of urbanization, shifts in dietary habits, and surge in sedentary lifestyles have all fueled this escalating epidemic. This crisis is further exacerbated by the concomitant worldwide rise in obesity [16].

Diabetes encompasses a group of metabolic disorders characterized by prolonged hyperglycemia due to deficits in insulin production, efficiency, or both. Persistent hyperglycemia, a defining trait of diabetes, is associated with long-term damage to various organs, particularly the eyes, kidneys, nerves, heart, and blood vessels, and their dysfunction and eventual failure [17].

People living with diabetes face a heightened risk of developing severe, long-term complications. These include retinopathy, a condition that may lead to vision loss; nephropathy, which can result in renal failure; and peripheral neuropathy, which escalates the risk of foot ulcers, amputations, and Charcot joints. Autonomic neuropathy, another possible side effect, can manifest in a variety of symptoms affecting the gastrointestinal, genitourinary, and cardiovascular systems, and lead to sexual dysfunction. Moreover, individuals with diabetes are more likely to experience atherosclerotic cardiovascular, peripheral arterial, and cerebrovascular diseases [17]. This highlights the multisystemic impact of diabetes and the critical need for effective management and prevention strategies.

HIV: ONE OF THE MOST PREVALENT AND DEADLY HUMAN DISEASES

According to estimates from the World Health Organization, the spread of HIV has been a significant global health issue, with at least 14,000 new infections reported daily in 2001 [18]. This virus can transmit in several ways. For instance, it can spread from mother to child during the perinatal period, through exposure to blood and blood products parenterally, and during sexual intercourse [19, 20].

Despite the critical role that antiretroviral therapy plays in managing the disease, importantly, HIV-infected patients under this treatment may still face an increased risk of numerous complications beyond acquired immune deficiency syndrome (AIDS) [21]. The consequences of the HIV epidemic are alarmingly severe, as reflected in the 2005 statistics: over 3 million individuals died from HIV-related complications, and nearly 5 million people became newly infected [22]. Therefore, the necessity of continuous research, prevention, and treatment strategies for HIV cannot be overstated.

COVID-19: ONE OF THE DEADLIEST HUMAN INFECTIOUS DISEASES

The coronavirus disease 2019 (COVID-19) pandemic has had a substantial impact on populations worldwide, both in industrialized and underdeveloped countries [23]. Severe acute respiratory syndrome coronavirus 2 (SARS-CoV-2), the virus responsible for the pandemic, spreads through droplet inhalation and direct physical contact with contaminated surfaces. Consequently, research facilities worldwide have been relentlessly pursuing the creation of effective vaccines [24].

In addition to the immediate risk of infection, SARS-CoV-2 is particularly damaging due to its significant effect on a range of human organs. This is due to its ability to engage with the angiotensin-converting enzyme 2 (ACE2) receptor [25]. As a result, individuals with preexisting conditions such as CVD, diabetes, obesity, asthma, chronic obstructive pulmonary disease, immunological deficiency, chronic renal impairment, and neurodegenerative diseases are facing higher mortality rates than those without such conditions.

Simultaneously, many developing nations are undergoing rapid economic and social development, leading to significant changes in lifestyle habits and dietary structures. These changes promote overnutrition and positive energy balance [26]. The quick pace of this nutritional transition is causing the coexistence of over- and undernutrition in many countries, leading to the dual burden of infectious and chronic diseases.

This fast-rising prevalence of risk factors, notably diabetes, hypertension, and obesity, especially among impoverished populations, is likely to lead to severe consequences. These are challenges that emerging countries may struggle to manage, primarily because of the lack of attention given to chronic ailments such as chronic kidney disease. This neglect is partly due to the focus of the global healthcare domain on infectious diseases and a lack of awareness [27].

In the quest to combat COVID-19, a wide range of potential therapeutic agents have been identified based on the transmission modes and genomic architecture of SARS-CoV-2 [28]. Although most of these drugs are still in the preclinical phase, they show considerable promise as anti-COVID-19 treatments.

In addition, gene therapy has been identified as a promising field of study, especially due to significant technical breakthroughs and effective treatments for specific diseases developed since the 1980s [29]. However, skepticism among some scientific communities has hindered progress in this area.

Despite major hurdles that need to be addressed before the widespread administration of immunotherapeutics to a diverse patient population, the introduction of novel drug delivery technologies may help overcome many of these challenges [30]. In particular, nanomedicine agents, including viral vectors, drug conjugates, lipid-based nanocarriers, polymer-based nanocarriers, and inorganic nanoparticles, have found extensive applications in clinical oncology [31]. Thus, the convergence of these emerging technologies and our understanding of disease mechanisms holds immense promise for the future of healthcare.

NEXT-GENERATION NANOMEDICINES AND COMBINATION THERAPIES: BRIDGING THE GAP IN GLOBAL HEALTH EQUITY

Materials existing at the nanoscale exhibit unique physicochemical interactions with their environment. As such, assessing their potential toxicity is fundamental both for regulatory purposes and to ensure the safer progression of nanomedicines [32]. The first generation of nanomedicine drugs, for example, pegylated liposomal doxorubicin and nab-paclitaxel, utilize passive targeting mechanisms [31].

However, as our understanding of nanoscale interactions expands, it becomes clear that the future of nanomedicine lies in developing the next generation of drugs. These will harness more sophisticated functionalities, establishing themselves as 'intelligent nanocarriers.' Second-generation nanomedicines are characterized by stimuli-responsive properties or active targeting vectors. This level of sophistication enhances drug targeting and boosts efficacy, paving the way for advancements in disease treatment at the nanoscale [31].

Furthermore, the potential of combination therapies is being explored in the biotechnological landscape. Unlike traditional monotherapies, combination therapies could leverage hitherto unidentified cellular and molecular mechanisms. Distinguishing between these therapeutic approaches is crucial, as it helps determine whether insights from monotherapy mechanisms can be applied to understand the functionality of combination treatments [33].

Indeed, the field of biotechnology is relentlessly seeking innovative approaches to expedite the development and dissemination of pharmaceuticals. These efforts are particularly targeted at addressing diseases predominantly affecting populations in developing nations. In so doing, the hope is to bridge the healthcare disparity between developed and developing countries, bringing us closer to global health equity [34].

Alzheimer's Disease and Dementia: Rising Global Prevalence and Innovative Genome Editing Approaches

According to the World Health Organization, Alzheimer's disease and other forms of dementia have emerged among the top 10 global killers [35]. Notably, Alzheimer's disease is increasingly becoming the most common cause of dementia [36]. As per death certificate statistics from 2018, 122,019 people were identified to have succumbed to Alzheimer's disease dementia, which translates to an average daily death count of 334.1 [37].

The global impact of dementia is projected to grow even more substantial in the future. By 2040, the number of people living with dementia is expected to reach 81.1 million, exhibiting a trend of doubling every 20 years [38].

This escalating prevalence of Alzheimer's disease has stimulated scientists to develop innovative tools and techniques for understanding and potentially treating this complex condition. One such technological advancement is genome editing, a technique that allows precise alterations such as the insertion or deletion of genes in cells or organisms. CRISPR-Cas9 is a genome editing technique that is being used in studies to modify specific DNA sequences associated with genetic disorders. This technique shows promise for applications in diseases such as Alzheimer's disease, with both *in vitro* and *in vivo* research underway [39].

Advanced Therapy Medicinal Products: Harnessing Gene Therapy for Disease Treatment

Gene therapy is a burgeoning field of study holding tremendous promise for treating or even curing ailments that are currently either untreatable or treated merely symptomatically [40]. Classed as an advanced therapy medicinal product (ATMP), gene therapy involves manipulating genetic material within human cells by introducing, removing, or altering genes.

Several distinct forms of gene therapies exist, each with unique potential. One of them is recombinant nucleic acid therapy, where hereditary material is synthetically produced *via* the fusion of DNA or RNA from various origins [41]. This form of therapy relies heavily on recombinant DNA technology, allowing researchers to manipulate and modify hereditary material to produce novel sequences not found in nature [42]. These therapies aim to cure or alleviate genetic diseases by introducing new genetic material into the patient's cells [42].

A closely related approach is DNA therapy, a gene therapy subset that uses recombinant DNA or nucleic acid to treat diseases caused by genetic mutations or abnormalities [43]. This method involves inserting a functional copy of a damaged gene into the patient's cells by employing various delivery methods such as viral vectors, plasmid DNA, and oligonucleotides [43].

In parallel, complementary DNA (cDNA), created from a messenger RNA (mRNA) prototype through the reverse transcription approach, is frequently used to examine gene expression and construct recombinant DNA for gene therapy [44, 45].

Moreover, the field also encompasses nucleic acid therapy, which administers nucleic acids—specifically DNA or RNA—as medicinal agents to manage illnesses [46]. This approach involves transmitting nucleic acids to cells to rectify genetic anomalies or regulate gene expression. Gene transfer, another subset of gene therapy, entails introducing genetic material into a person's cells to treat or prevent diseases [47]. Various strategies exist for accomplishing this, such as directly injecting the genetic material into the cells or using viruses to deliver it.

Closely linked to gene transfer is the concept of "virus delivery," which involves using viruses as vectors for transporting genetic material into cells [48]. This method is widely employed in gene therapy. RNA therapy utilizes RNA molecules as therapeutic agents to treat a range of diseases [49]. It involves deploying small interfering RNA (siRNA) to suppress specific disease-implicated genes or mRNA to generate therapeutic proteins within the organism. Relatedly, cancer immunotherapy, a type of gene therapy, uses the body's immune system to combat cancerous cells [50]. Similarly, tumor vaccines prompt the immune system to identify and attack cancerous cells [51].

The use of plasmid DNA, self-replicating circular DNA molecules distinct from chromosomal DNA, and oligonucleotides, abbreviated nucleotide sequences, is also common in gene therapy for delivering therapeutic genes to specific cells or for selectively targeting genes or gene products [52, 53].

In the broader context, the term "genetically modified organisms" refers to organisms whose genetic constitution has been unnaturally altered through techniques not involving natural mating or recombination [54]. Moreover, "genetically modified cells," which have been genetically engineered to introduce or remove specific genes or modify gene expression, find frequent use in gene therapy and somatic cell therapy [55].

The four major categories of ATMPs are as follows: tissue-engineered products (TEPs), somatic cell therapy medicinal products (sCTMPs), gene therapy medicinal products (GTMPs), and combination products (tissues or cells associated with a device) [55]. As these therapeutic modalities continue to develop, they are expected to yield significant health benefits and influence pharmaceutical budgets. The aim is to continually improve the profiles of ATMPs under development and deliberate the future implications for market access.

The dynamic field of gene therapy, in all its myriad forms, holds immense promise for the future of medicine. Through the introduction, removal, or alteration of genetic material, these advanced therapies seek to treat and potentially cure a range of diseases for which treatments or cures have hitherto been elusive. We are in an exciting era as we see science move from simply managing symptoms to correcting the very genetic fabric responsible for disease. As we continue to unravel the complexities of the human genome and as our understanding of genetic manipulation grows, the possibilities seem almost limitless [55]. A related figure illustrating some examples of biotechnology for disease treatment is shown in Fig. (**2**).

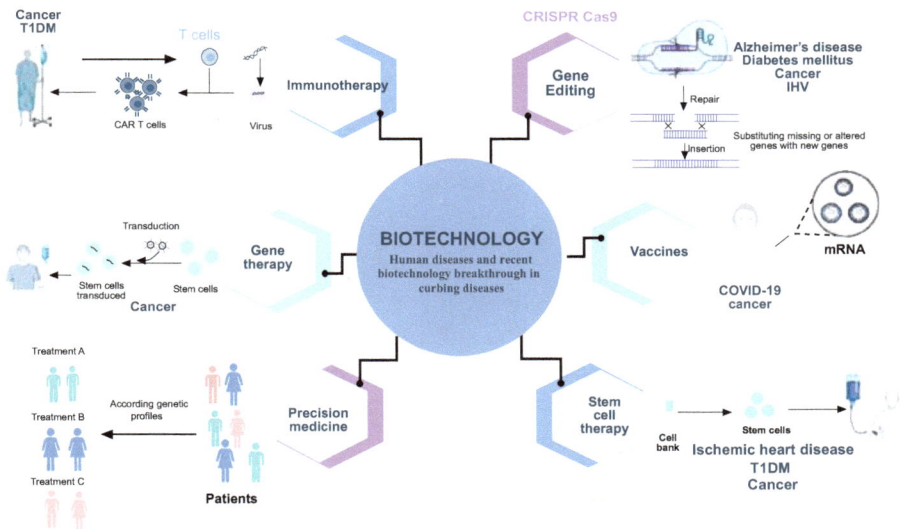

Fig. (2). Recent biotechnology breakthrough in curbing diseases. Adapted from "Cell and Gene Therapy" by BioRender.com.

Immunotherapy: A Paradigm Shift in Cancer Treatment

Cancer immunotherapy, an innovative therapeutic strategy, harnesses the power of the immune system to elicit an antitumor response. This approach, which has transitioned from a promising possibility to a solid clinical reality over the past decade, essentially has two main branches: active therapies and passive therapeutics [56].

Passive immunotherapy entails approaches that do not directly activate the patient's immune system but rather introduce 'trained' agents to combat cancer. This subfield of immunotherapy is exemplified by techniques such as the injection of monoclonal antibodies and cytokines and the adoptive cell transfer of immune cells that have been 'trained' *ex vivo* to recognize and destroy cancer cells [56].

On the other hand, active immunotherapy aims to directly stimulate the patient's own immune system to fight the cancer. Examples of active immunotherapy include anticancer vaccines (including components such as peptides and dendritic cells (DCs)), allogeneic whole-cell vaccines, immune checkpoint inhibitors, and oncolytic viruses. These treatments operate by engaging and activating the patient's immune responses to target and eliminate tumor cells [56].

DCs play a critical role in active immunotherapy, functioning as messengers between the innate and adaptive immune systems. DC vaccines, for instance, have been developed to harness the ability of DCs to induce immunity, effectively teaching the immune system to identify and attack cancer cells [56].

In addition to these established strategies, the field of cancer immunotherapy is also marked by the continuous exploration of novel approaches aimed at strengthening anticancer immune responses. These emerging techniques promise to further expand our arsenal in the fight against cancer, ultimately offering renewed hope for patients worldwide [56].

Precision Medicine: The Future of Individualized Treatment Strategies

Precision medicine is an approach in which individual variability in prevention and treatment strategies is considered, although the concept itself is not new. Blood transfusions, for instance, have been managed using blood type for nearly a century. However, recent advancements in large-scale biologic databases, robust patient-characterization techniques, and mobile health technology have brought us closer to the broad application of this concept [57].

Our enhanced ability to sequence the human genome and our advancements in proteomics, metabolomics, genomics, and various cellular assays have spurred a revolution in precision medicine [57]. These advancements allow us to understand a patient's unique genetic makeup and physiological state more accurately than ever before, thereby facilitating more targeted and effective treatment strategies.

Moreover, synthetic biology strategies applied in the field of cell engineering present new horizons for precision medicine. Specifically, they could lead to the development of 'smart' T-cell precision therapies that are designed to recognize and target the unique characteristics of individual tumors [58].

By combining these cutting-edge techniques with precision informatics, it may become feasible to create truly individualized therapies meticulously tailored to each patient's specific needs [58]. This will usher in a new era of healthcare where treatments are not one-size-fits-all but are customized to each patient's unique physiological and genetic profile.

The existing pillars of oncology, such as prevention, diagnostics, specific screening techniques, and effective therapies, are set to be enhanced and refined by the emphasis of precision medicine in cancer treatment. This approach is not intended to replace the current standards but rather to complement them, acting as a robust framework to expedite the adoption of precision medicine in various disciplines [57]. Thus, the potential of precision medicine goes beyond oncology, promising a revolution in the broader field of healthcare by introducing individualized treatment strategies.

Harnessing Biotechnology in the Battle Against Infectious Diseases

Biotechnology has proven to be a powerful tool in the fight against HIV and other infectious diseases. Pioneering gene-editing tools such as CRISPR have allowed researchers to identify and study genomic sequences unique to HIV, providing new avenues for therapeutic intervention [59].

CRISPR/Cas9, an antiviral gene-editing and gene-modulating tool, can be used therapeutically to target the provirus. It does this by coupling the Cas9 nuclease with one or more small guide RNAs (sgRNAs), facilitating the excision of the integrated viral genome. In addition, a modified, nuclease-deficient Cas9 coupled with transcription-activating domains can selectively stimulate proviral gene expression, potentially enabling the elimination of latent reservoirs [59].

These technologies are not limited to directly targeting the virus alone. They can also be used to interrupt the interaction of the virus with host cells. One such application is targeting host dependence factors such as the coreceptor CCR5, a strategy that can prevent the virus from gaining entry into the cells [59].

Biotechnology also plays a crucial role in enhancing the immune response in HIV-infected individuals who are also battling cancer. The administration of antibodies against immune checkpoints, which serve to regulate the immune system and prevent its overactivation, has shown promising results. This approach has been demonstrated to be effective and without severe side effects, significantly bolstering the immune response against cancer in HIV-infected individuals [60].

The use of antiretroviral therapy (ART), a combination treatment using several antiretroviral drugs, has greatly improved the lives of many people living with HIV. A significant reduction in risks associated with the onset of AIDS and HIV-related inflammatory events is observed in patients who begin ART early in their infection [61].

Despite these strides, creating an effective HIV-1 vaccine presents a unique challenge. The complexity of HIV-1 and its genetic diversity, rapid mutation rate, and ability to evade immune responses demand a far deeper understanding of virus-host interactions than has been required or achieved for any previous vaccination efforts. This makes the development of an HIV-1 vaccine a multifaceted and challenging process [62].

Biotechnology thus remains an essential ally in our ongoing efforts to understand, prevent, and treat HIV and other infectious diseases. Continued developments in the field of biotechnology will be pivotal in enabling us to surmount the challenges that lay ahead in this field.

Utilizing Biotechnology in the Development of Treatments for Chronic Diseases Such as Cancer

Significant strides in the development of treatments for chronic diseases, particularly cancer, have been made in the field of biotechnology. As a direct result of these advancements, oncogenes, which are genes that have the potential to cause cancer, are being used as novel therapeutic targets. These targets can be utilized for cancer detection, prognosis, and treatment, which opens up an array of possibilities for the future of oncology [63].

One of the most significant recent developments in cancer treatment has been the advent of combination immunotherapies. These innovative treatments have shown unprecedented efficacy in responding to patients, offering a new level of hope in the fight against cancer [64].

Moreover, the application of combinatorial strategies in targeting cancer represents a highly promising direction. By integrating various treatment approaches, these strategies aim to enhance the efficacy of targeted anticancer drug delivery. The overall success rate of cancer treatment protocols could potentially be augmented through this approach, which harmonizes different therapeutic interventions for a synergistic effect [65].

Biotechnology is playing a vital role in reshaping our approach to chronic diseases, particularly cancer. Identifying novel therapeutic targets, enhancing treatment efficacy through combination therapies, and developing advanced drug delivery strategies will pave the way for a future where cancer is a less formidable adversary.

Leveraging Biotechnology in the Development of Treatments for Diabetes Mellitus

In the ever-evolving field of biotechnology, researchers are developing innovative treatments for chronic diseases such as diabetes mellitus. One promising approach involves using gene-editing technologies to target specific genes related to diabetes. These cutting-edge methods are aimed at improving glucose tolerance and boosting insulin production, key factors in managing diabetes [66].

A specific genetic variant, when targeted, may aid a unique subset of individuals diagnosed with type 2 diabetes. Researchers propose a treatment involving an antagonist, a drug designed to inhibit a particular molecule or receptor from functioning. This proposal has a direct correlation with the physiological process of insulin secretion, which is hampered in type 2 diabetes as the body becomes resistant to insulin, leading to elevated blood sugar levels. The use of an antagonist to target this genetic variant could potentially restore insulin secretion in a subgroup of type 2 diabetes patients, opening up a new avenue for personalized diabetes treatment [66].

In the realm of type 1 diabetes mellitus (T1DM), a convergent approach involving immunotherapy, stem cell-mediated generation of pancreatic beta-cells, and bioengineering holds great promise. In T1DM, the immune system attacks insulin-producing β-cells, leading to increased blood sugar levels. Immunotherapy aims to manipulate the immune system to either suppress these autoreactive immune cells or stimulate regulatory immune cells that can prevent this attack [67].

Stem cell therapy, another facet of this approach, entails using undifferentiated cells that can morph into different cell types. The aim here is to use these cells to generate new β-cells or islets capable of secreting insulin, thus replacing those lost in T1DM patients and restoring insulin production [67].

The third pillar of this tri-fold approach involves bioengineering, a field dedicated to creating and designing materials or objects that can interact with biological entities. In relation to T1DM, bioengineering can be used to design a protective barrier around newly transplanted β-cells or islets, preventing immune attack and rejection [67].

The convergence of these three methodologies — immunotherapy, stem cell therapy and bioengineering — has the potential to bring about a long-lasting treatment or even a medical cure for T1DM. Currently managed with insulin injections or pumps, patients with T1DM could see significant improvements through these pioneering techniques [67].

NAVIGATING CHALLENGES OF BIOTECHNOLOGY IN DRUG DEVELOPMENT: COST, ACCESS, AND ETHICAL CONCERNS

Various challenges are encountered in the field of drug development, and these challenges must be addressed to fully realize the potential of biotechnology. These challenges span various dimensions, including cost implications, access issues, and ethical concerns, all of which impact the path to successful drug development and application [68].

A key element of drug development involves leveraging a variety of techniques, such as computational methods, in silico technology, and computer-based predictive models. These methods promise to significantly curtail the exorbitant costs associated with drug development. However, the potential of these technologies has yet to be fully defined, and it remains uncertain whether they will truly facilitate the much-needed reduction in drug development costs [68].

Pharmacogenomic studies hold significant potential for cost reduction in preclinical and clinical phase III testing. By enabling early elimination of certain drugs and identifying a more selective target population for the remaining drugs, pharmacogenomic studies can streamline the drug development process. Indeed, tailoring drug therapy to individual genetic profiles can bring about a significant decrease in adverse outcomes, emphasizing the value of precision medicine. However, this approach necessitates careful consideration of ethical implications, including the potential for smaller-than-anticipated markets for certain drugs [68].

Moreover, the size of phase II-III clinical trials, which typically involve just 2,000–3,000 participants, constrains the ability to identify rarer drug side effects. With this limited sample size, only undesirable effects with a frequency of 1 in 1,000–1 in 1,500 are likely to be detected. The paradox lies in the fact that the rarer side effects, often missed in these studies, are frequently the most severe.

The occurrence of these side effects can prompt the swift withdrawal of a drug from the market, even after millions have been invested in its development [68].

In light of these complexities, the future of biotechnology in drug development is a balancing act of cost, efficacy, safety, access, and ethical considerations. The path forward will necessitate careful navigation and a comprehensive understanding of these various elements to fully harness the potential of biotechnology in this crucial field [68].

ROLE OF MACHINE LEARNING AND ARTIFICIAL INTELLIGENCE IN THE ADVANCEMENT OF BIOLOGICS, CELL TREATMENTS, AND DRUG DISCOVERY

In recent years, the application of machine learning (ML) and artificial intelligence (AI) has revolutionized the process of drug development, enhancing both its effectiveness and accuracy. This section presents a comprehensive analysis of the recent research trends in AI-assisted drug discovery, touching upon key stages including target identification; hit discovery; absorption, distribution, metabolism, excretion, and toxicity (ADMET) prediction; lead optimization; and drug repositioning [69].

First, AI and ML have proven to be invaluable tools for the identification of potential therapeutic targets. By analyzing large volumes of biological data, these methods facilitate the discovery of patterns and relationships among genes, proteins, and diseases, thus opening the way to the identification of new therapeutic targets [70]. This technology is also being utilized in hit identification, where AI and ML techniques are deployed to screen vast databases of chemical compounds to identify potential drug candidates that are likely effective against specific targets. This innovative approach dramatically reduces the time and cost associated with traditional screening methods, accelerating the discovery of new drug candidates [71].

A pivotal aspect of drug development involves predicting the ADMET characteristics of potential drug candidates. Here, AI and ML techniques have also shown promise, enabling the identification of compounds that are likely to be safe and effective [72]. In the realm of lead optimization, AI and ML techniques have been deployed to enhance the likelihood of new drug candidates succeeding in clinical trials by optimizing the potency, selectivity, and pharmacokinetic properties of these candidates [73]. Last, AI and ML techniques are being increasingly applied in the field of drug repositioning. By analyzing large volumes of data, these techniques can identify new uses for existing drugs based on observed similarities between diseases [74].

CONCLUSION

The utilization of biotechnology and AI in the sphere of medicine has exhibited tremendous potential, especially in the creation of new and innovative therapeutic interventions for diverse ailments. As extensively investigated, their capabilities in areas such as cancer immunotherapy, precision medicine, and treatment for chronic afflictions such as diabetes and HIV are profound and substantial.

Cancer immunotherapy has advanced from a mere possibility to a robust clinical actuality over the past decade. The potential for precision medicine, amplified by strides in large-scale biological databases and patient characterization techniques, is poised to improve current pillars of oncology, including prevention, diagnostic and treatment strategies. Furthermore, biotechnology has demonstrated noteworthy contributions in comprehending and treating infectious diseases such as HIV and chronic diseases such as cancer and diabetes.

However, despite these encouraging developments, it is imperative to acknowledge the challenges in the field of biotechnology in drug development. Issues concerning cost, accessibility, and ethical considerations are paramount. It is evident that novel technologies have the potential to exponentially reduce the costs associated with drug development, but their precise application and impact are yet to be thoroughly defined. Additionally, there are significant ethical considerations that must be prudently managed throughout the development process.

Concurrently, AI and ML have revolutionized the drug discovery process by hastening target identification, hit discovery, ADMET prediction, lead optimization, and drug repositioning, indicating that these advanced technologies are not only streamlining the discovery process but also reducing costs and optimizing efficacy.

In conclusion, the amalgamation of biotechnology, precision medicine, and AI presents immense promise in the medical field, with substantial implications for the treatment of chronic and infectious diseases. While challenges remain, the potential benefits are significant. As we continue to unlock the full potential of these techniques, we can anticipate a future where treatment strategies are personalized, diseases are diagnosed earlier, and therapeutic interventions are more effective. However, it is crucial to navigate ethical considerations and ensure equitable access to these advancements to fully realize the benefits of this progress.

ACKNOWLEDGMENTS

This work was supported by the UNAM Postdoctoral Program (POSDOC) and Universidad Nacional Autónoma de México-DGAPA-PAPIIT-UNAM (IN-212623).

REFERENCES

[1]   Farooq Z, Rashid S, Mahmood S, Mahmood A, Anwar M. The Advent of Medical Biotechnology. In: Anwar M, Rather RA, Farooq Z, Eds. Fundamentals and Advances in Medical Biotechnology Switzerland: Springer Cham 2022; pp.1-20.
      [http://dx.doi.org/10.1007/978-3-030-98554-7_1]

[2]   Liao C, Xiao S, Wang X. Bench-to-bedside: Translational development landscape of biotechnology in healthcare. Health Sci Rev 2023; 7: 100097.

[3]   Wong E, Bertin N, Hebrard M, *et al.* The Singapore national precision medicine strategy. Nat Genet 2023; 55(2): 178-86.
[http://dx.doi.org/10.1038/s41588-022-01274-x] [PMID: 36658435]

[4]   Naithani N, Sinha S, Misra P, Vasudevan B, Sahu R. Precision medicine: Concept and tools. Med J Armed Forces India 2021; 77(3): 249-57.
[http://dx.doi.org/10.1016/j.mjafi.2021.06.021] [PMID: 34305276]

[5]   Yu J, Li T, Zhu J. Gene therapy strategies targeting aging-related diseases. Aging Dis 2023; 14(2): 398-417.
[PMID: 37008065]

[6]   Guo N, Liu JB, Li W, Ma YS, Fu D. The power and the promise of CRISPR/Cas9 genome editing for clinical application with gene therapy. J Adv Res 2022; 40: 135-52.
[http://dx.doi.org/10.1016/j.jare.2021.11.018] [PMID: 36100322]

[7]   Esfahani K, Roudaia L, Buhlaiga N, Del Rincon SV, Papneja N, Miller WH Jr. A review of cancer immunotherapy: from the past, to the present, to the future. Curr Oncol 2020; 27 (S2): 87-97.
[http://dx.doi.org/10.3747/co.27.5223] [PMID: 32368178]

[8]   Strianese O, Rizzo F, Ciccarelli M, *et al.* Precision and personalized medicine: How genomic approach improves the management of cardiovascular and neurodegenerative disease. Genes (Basel) 2020; 11(7): 747.
[http://dx.doi.org/10.3390/genes11070747] [PMID: 32640513]

[9]   Bai J, Cui J, Shi F, Yu C. Global epidemiological patterns in the burden of main non-communicable diseases, 1990–2019: Relationships with Socio-demographic index. Int J Public Health 2023; 68: 1605502.
[http://dx.doi.org/10.3389/ijph.2023.1605502] [PMID: 36726528]

[10]   Balakumar P, Maung-U K, Jagadeesh G. Prevalence and prevention of cardiovascular disease and diabetes mellitus. Pharmacol Res 2016; 113(Pt A): 600-9.
[http://dx.doi.org/10.1016/j.phrs.2016.09.040] [PMID: 27697647]

[11]   Buddai SK, Toulokhonova L, Bergum PW, Vlasuk GP, Krishnaswamy S. Nematode anticoagulant protein c2 reveals a site on factor Xa that is important for macromolecular substrate binding to human prothrombinase. J Biol Chem 2002; 277(29): 26689-98.
[http://dx.doi.org/10.1074/jbc.M202507200] [PMID: 12011050]

[12]   Segers VFM, Lee RT. Stem-cell therapy for cardiac disease. Nature 2008; 451: 937-42.
[http://dx.doi.org/10.1038/nature06800] [PMID: 18288183]

[13]   Carmeliet P. Angiogenesis in health and disease. Nat Med 2003; 9(6): 653-60.
[http://dx.doi.org/10.1038/nm0603-653] [PMID: 12778163]

[14]   Ferrara N, Kerbel RS. Angiogenesis as a therapeutic target. Nature 2005; 438: 967-74.
[http://dx.doi.org/10.1038/nature04483] [PMID: 16355214]

[15]   de Martel C, Ferlay J, Franceschi S, *et al.* Global burden of cancers attributable to infections in 2008: A review and synthetic analysis. Lancet Oncol 2012; 13(6): 607-15.
[http://dx.doi.org/10.1016/S1470-2045(12)70137-7] [PMID: 22575588]

[16]   Sung H, Ferlay J, Siegel RL, *et al.* Global cancer statistics 2020: GLOBOCAN estimates of incidence and mortality worldwide for 36 cancers in 185 countries. CA Cancer J Clin 2021; 71(3): 209-49.
[http://dx.doi.org/10.3322/caac.21660] [PMID: 33538338]

[17]   American Diabetes Association. Diagnosis and classification of diabetes mellitus. Diabetes Care 2014; 37(1): S81-90.
[PMID: 24357215]

[18]   Mwau M, McMichael AJ. A review of vaccines for HIV prevention. J Gene Med 2003; 5(1): 3-10.
[http://dx.doi.org/10.1002/jgm.343] [PMID: 12516046]

[19]   Kapoor A, Kapoor A, Vani SN. Prevention of mother to child transmission of HIV. Indian J Pediatr 2004; 71(3): 247-51.
[http://dx.doi.org/10.1007/BF02724278] [PMID: 15080412]

[20]   Jaffe HW, Lifson AR. Acquisition and transmission of HIV. Infect Dis Clin North Am 1988; 2(2): 299-306.
[http://dx.doi.org/10.1016/S0891-5520(20)30184-7] [PMID: 3060517]

[21]   Deeks SG, Phillips AN. HIV infection, antiretroviral treatment, ageing, and non-AIDS related morbidity. BMJ 2009; 338: a3172.
[http://dx.doi.org/10.1136/bmj.a3172] [PMID: 19171560]

[22]   Shors AR. The global epidemiology of HIV/AIDS. Dermatol Clin 2006; 24(4): 413-20.
[http://dx.doi.org/10.1016/j.det.2006.06.004] [PMID: 17010771]

[23]   Elsevier's Novel Coronavirus Information Center 2020. Available from: https://beta.elsevier.com/connect/coronavirus-information-center?trial=true

[24]   Atzrodt CL, Maknojia I, McCarthy RDP, *et al.* A Guide to COVID 19: A global pandemic caused by the novel coronavirus SARS CoV2. FEBS J 2020; 287(17): 3633-50.
[http://dx.doi.org/10.1111/febs.15375] [PMID: 32446285]

[25]   Fitero A, Bungau SG, Tit DM, *et al.* Comorbidities, associated diseases, and risk assessment in COVID-19 : A systematic review. Int J Clin Pract 2022; 2022: 1-24.
[http://dx.doi.org/10.1155/2022/1571826] [PMID: 36406478]

[26]   Hu FB. Globalization of Diabetes. Diabetes Care 2011; 34(6): 1249-57.
[http://dx.doi.org/10.2337/dc11-0442] [PMID: 21617109]

[27]   Bommer C, Heesemann E, Sagalova V, *et al.* The global economic burden of diabetes in adults aged 20–79 years: A cost of illness study. Lancet Diabetes Endocrinol 2017; 5(6): 423-30.
[http://dx.doi.org/10.1016/S2213-8587(17)30097-9] [PMID: 28456416]

[28]   Mulaw Belete T. An up to date overview of therapeutic agents for the treatment of COVID-19 disease. Clin Pharmacol 2020; 12: 203-12.
[http://dx.doi.org/10.2147/CPAA.S284809] [PMID: 33363416]

[29]   Kay MA. State of the art gene-based therapies: the road ahead. Nat Rev Genet 2011; 12(5): 316-28.
[http://dx.doi.org/10.1038/nrg2971] [PMID: 21468099]

[30]   Riley RS, June CH, Langer R, Mitchell MJ. Delivery technologies for cancer immunotherapy. Nat Rev Drug Discov 2019; 18(3): 175-96.
[http://dx.doi.org/10.1038/s41573-018-0006-z] [PMID: 30622344]

[31]   Wicki A, Witzigmann D, Balasubramanian V, Huwyler J. Nanomedicine in cancer therapy: Challenges, opportunities, and clinical applications. J Control Release 2015; 200: 138-57.
[http://dx.doi.org/10.1016/j.jconrel.2014.12.030] [PMID: 25545217]

[32]   Singh AV, Rosenkranz D, Ansari MHD, *et al.* Artificial intelligence and machine learning empower advanced biomedical material design to toxicity prediction. Adv Intell Syst 2020; 2(12): 2000084.
[http://dx.doi.org/10.1002/aisy.202000084]

[33]   Wei SC, Duffy CR, Allison JP. Fundamental mechanisms of immune checkpoint blockade therapy. Cancer Discov 2018; 8(9): 1069-86.
[http://dx.doi.org/10.1158/2159-8290.CD-18-0367] [PMID: 30115704]

[34]   Richard Gold E, Morin JF. Promising trends in access to medicines. Glob Policy 2012; 3(2): 231-7.
[http://dx.doi.org/10.1111/j.1758-5899.2011.00110.x] [PMID: 32336993]

[35]   GBD 2019 Collaborators. Global mortality from dementia: Application of a new method and results

from the Global Burden of Disease Study 2019. Alzheimers Dement 2021; 7(1): e12200.
[http://dx.doi.org/10.1002/trc2.12200] [PMID: 34337138]

[36]   Lane CA, Hardy J, Schott JM. Alzheimer's disease. Eur J Neurol 2018; 25(1): 59-70.
        [http://dx.doi.org/10.1111/ene.13439] [PMID: 28872215]

[37]   Cummings J, Lee G, Zhong K, Fonseca J, Taghva K. Alzheimer's disease drug development pipeline:
        2021. Alzheimers Dement 2021; 7(1): e12179.
        [http://dx.doi.org/10.1002/trc2.12179] [PMID: 34095440]

[38]   Ferri CP, Prince M, Brayne C, *et al.* Global prevalence of dementia: A Delphi consensus study. Lancet
        2005; 366: 2112-7.
        [http://dx.doi.org/10.1016/S0140-6736(05)67889-0] [PMID: 16360788]

[39]   Mani I. CRISPR-Cas9 for treating hereditary diseases. In: Singh V, Eds. Progress in Molecular
        Biology and Translational Science. San Diego, CA, Estados Unidos de América: Elsevier; 2021; pp.
        165–83.

[40]   Gonçalves GAR, Paiva RMA. Gene therapy: Advances, challenges and perspectives. Einstein 2017;
        15(3): 369-75.
        [http://dx.doi.org/10.1590/s1679-45082017rb4024] [PMID: 29091160]

[41]   Nagy PD, Simon AE. New insights into the mechanisms of RNA recombination. Virology 1997;
        235(1): 1-9.
        [http://dx.doi.org/10.1006/viro.1997.8681] [PMID: 9300032]

[42]   Iglesias-López C, Agustí A, Obach M, Vallano A. Regulatory framework for advanced therapy
        medicinal products in Europe and United States. Front Pharmacol 2019; 10: 921.
        [http://dx.doi.org/10.3389/fphar.2019.00921] [PMID: 31543814]

[43]   Marangi M, Pistritto G. Innovative therapeutic strategies for cystic fibrosis: Moving forward to
        CRISPR technique. Front Pharmacol 2018; 9: 396.
        [http://dx.doi.org/10.3389/fphar.2018.00396] [PMID: 29731717]

[44]   Liang P, Pardee AB. Analysing differential gene expression in cancer. Nat Rev Cancer 2003; 3(11):
        869-76.
        [http://dx.doi.org/10.1038/nrc1214] [PMID: 14668817]

[45]   Glick BR, Patten CL. Molecular Biotechnology: Principles and applications of Recombinant DNA.
        John Wiley & Sons 2022; pp. 876.

[46]   Kulkarni JA, Witzigmann D, Thomson SB, *et al.* The current landscape of nucleic acid therapeutics.
        Nat Nanotechnol 2021; 16(6): 630-43.
        [http://dx.doi.org/10.1038/s41565-021-00898-0] [PMID: 34059811]

[47]   Sayed N, Allawadhi P, Khurana A, *et al.* Gene therapy: Comprehensive overview and therapeutic
        applications. Life Sci 2022; 294(120375): 120375.
        [http://dx.doi.org/10.1016/j.lfs.2022.120375] [PMID: 35123997]

[48]   Giacca M, Zacchigna S. Virus mediated gene delivery for human gene therapy. J Control Release
        2012; 161(2): 377-88.
        [http://dx.doi.org/10.1016/j.jconrel.2012.04.008] [PMID: 22516095]

[49]   Kim YK. RNA therapy: Rich history, various applications and unlimited future prospects. Exp Mol
        Med 2022; 54(4): 455-65.
        [http://dx.doi.org/10.1038/s12276-022-00757-5] [PMID: 35440755]

[50]   Oiseth SJ, Aziz MS. Cancer immunotherapy: A brief review of the history, possibilities, and
        challenges ahead. J Cancer Metastasis Treat 2017; 3(10): 250-61.
        [http://dx.doi.org/10.20517/2394-4722.2017.41]

[51]   Ebben JD, Rocque BG, Kuo JS. Tumour vaccine approaches for CNS malignancies: Progress to date.
        Drugs 2009; 69(3): 241-9.

[http://dx.doi.org/10.2165/00003495-200969030-00001] [PMID: 19275269]

[52]   Sousa F, Passarinha L, Queiroz JA. Biomedical application of plasmid DNA in gene therapy: A new challenge for chromatography. Biotechnol Genet Eng Rev 2009; 26(1): 83-116.
[http://dx.doi.org/10.5661/bger-26-83] [PMID: 21415877]

[53]   Crooke ST. Therapeutic applications of oligonucleotides. Biotechnology 1992; 10(8): 882-6.
[PMID: 1280444]

[54]   Kaestner L, Scholz A, Lipp P. Conceptual and technical aspects of transfection and gene delivery. Bioorg Med Chem Lett 2015; 25(6): 1171-6.
[http://dx.doi.org/10.1016/j.bmcl.2015.01.018] [PMID: 25677659]

[55]   Hanna E, Rémuzat C, Auquier P, Toumi M. Advanced therapy medicinal products: Current and future perspectives. J Mark Access Health Policy 2016; 4(1): 31036.
[http://dx.doi.org/10.3402/jmahp.v4.31036] [PMID: 27123193]

[56]   Papaioannou NE, Beniata OV, Vitsos P, Tsitsilonis O, Samara P. Harnessing the immune system to improve cancer therapy. Ann Transl Med 2016; 4(14): 261.
[http://dx.doi.org/10.21037/atm.2016.04.01] [PMID: 27563648]

[57]   Collins FS, Varmus H. A new initiative on precision medicine. N Engl J Med 2015; 372(9): 793-5.
[http://dx.doi.org/10.1056/NEJMp1500523] [PMID: 25635347]

[58]   Lim WA, June CH. The principles of engineering immune cells to treat cancer. Cell 2017; 168(4): 724-40.
[http://dx.doi.org/10.1016/j.cell.2017.01.016] [PMID: 28187291]

[59]   Saayman S, Ali SA, Morris KV, Weinberg MS. The therapeutic application of CRISPR/Cas9 technologies for HIV. Expert Opin Biol Ther 2015; 15(6): 819-30.
[http://dx.doi.org/10.1517/14712598.2015.1036736] [PMID: 25865334]

[60]   Gonzalez-Cao M, Martinez-Picado J, Karachaliou N, Rosell R, Meyerhans A. Cancer immunotherapy of patients with HIV infection. Clin Transl Oncol 2019; 21(6): 713-20.
[http://dx.doi.org/10.1007/s12094-018-1981-6] [PMID: 30446984]

[61]   Routy JP, Mehraj V, Cao W. HIV immunotherapy comes of age: Implications for prevention, treatment and cure. Expert Rev Clin Immunol 2016; 12(2): 91-4.
[http://dx.doi.org/10.1586/1744666X.2016.1112269] [PMID: 26629806]

[62]   Haynes BF, Mascola JR. The quest for an antibody based HIV vaccine. Immunol Rev 2017; 275(1): 5-10.
[http://dx.doi.org/10.1111/imr.12517] [PMID: 28133795]

[63]   Huber BE. Therapeutic opportunities involving cellular oncogenes: Novel approaches fostered by biotechnology. FASEB J 1989; 3(1): 5-13.
[http://dx.doi.org/10.1096/fasebj.3.1.2642869] [PMID: 2642869]

[64]   Scheetz L, Park KS, Li Q, *et al.* Engineering patient specific cancer immunotherapies. Nat Biomed Eng 2019; 3(10): 768-82.
[http://dx.doi.org/10.1038/s41551-019-0436-x] [PMID: 31406259]

[65]   Alexander-Bryant AA, Vanden Berg-Foels WS, Wen X. Bioengineering strategies for designing targeted cancer therapies. In: Tew KD, Fisher PB, Eds. Advances in Cancer Research. San Diego, CA, Estados Unidos de América: Elsevier 2013; pp. 1–59.
[http://dx.doi.org/10.1016/B978-0-12-407173-5.00002-9]

[66]   Ostenson C. Type 2 Diabetes: Genotype-Based Therapy. 2014; 6: 8–10. Ostenson CG. Type 2 diabetes: Genotype-based therapy. Sci Transl Med 2014; 6(257): 257fs39.
[PMID: 25298318]

[67]   Kopan C, Tucker T, Alexander M, Mohammadi MR, Pone EJ, Lakey JRT. Approaches in immunotherapy, regenerative medicine, and bioengineering for type 1 diabetes. Front Immunol 2018;

9: 1354.
[http://dx.doi.org/10.3389/fimmu.2018.01354] [PMID: 29963051]

[68]　Preziosi P. Science, pharmacoeconomics and ethics in drug R&D: A sustainable future scenario. Nat Rev Drug Discov 2004; 3(6): 521-6.
[http://dx.doi.org/10.1038/nrd1418] [PMID: 15173841]

[69]　Kim H, Kim E, Lee I, Bae B, Park M, Nam H. Artificial intelligence in drug discovery: A comprehensive review of data-driven and machine learning approaches. Biotechnol Bioprocess Eng; BBE 2020; 25(6): 895-930.
[http://dx.doi.org/10.1007/s12257-020-0049-y] [PMID: 33437151]

[70]　Carpenter KA, Huang X. Machine learning-based virtual screening and its applications to alzheimer's drug discovery: A review. Curr Pharm Des 2018; 24(28): 3347-58.
[http://dx.doi.org/10.2174/1381612824666180607124038] [PMID: 29879881]

[71]　Smith JS, Roitberg AE, Isayev O. Transforming computational drug discovery with machine learning and AI. ACS Med Chem Lett 2018; 9(11): 1065-9.
[http://dx.doi.org/10.1021/acsmedchemlett.8b00437] [PMID: 30429945]

[72]　Kumar A, Kini SG, Rathi E. A recent appraisal of artificial intelligence and in silico ADMET prediction in the early stages of drug discovery. Mini Rev Med Chem 2021; 21(18): 2788-800.
[http://dx.doi.org/10.2174/1389557521666210401091147] [PMID: 33797376]

[73]　Dobchev D, Pillai G, Karelson M. *In silico* machine learning methods in drug development. Curr Top Med Chem 2014; 14(16): 1913-22.
[http://dx.doi.org/10.2174/1568026614666140929124203] [PMID: 25262800]

[74]　Tanoli Z, Vähä-Koskela M, Aittokallio T. Artificial intelligence, machine learning, and drug repurposing in cancer. Expert Opin Drug Discov 2021; 16(9): 977-89.
[http://dx.doi.org/10.1080/17460441.2021.1883585] [PMID: 33543671]

CHAPTER 9

Exploring the Intersection of Omics Technologies and Biotechnology in Drug Interaction Studies

Israel Valencia Quiroz[1,*]

[1] *Phytochemistry Laboratory, UBIPRO, Superior Studies Faculty (FES)-Iztacala, National Autonomous University of Mexico (UNAM), Tlalnepantla de Baz, México State, 54090, México*

Abstract: The integration of omics tools with biotechnology has led to a paradigm shift in our comprehension of drug interactions, providing profound insights into the molecular mechanisms underlying these interactions. We explore the crucial functions of genomes, transcriptomics, proteomics, and metabolomics in this chapter to decode pharmacological interactions at various molecular levels. Notably, significant emphasis is placed on the application of omics tools in areas such as high-throughput screening for unveiling novel drug targets, personalized medicine, pharmacogenomics, understanding drug-drug and drug-metabolite interactions, drug repurposing, polypharmacology, and systems biology. Furthermore, the paper explores the potential of integrating omics data with computational approaches to study complex biological networks, highlighting the instrumental role of microbial biotechnology in drug interactions. Importantly, alongside these advancements, there is also an in-depth discussion of the ethical, legal, and societal ramifications of the use of omics technologies in biotechnology. Moreover, the text presents an in-depth examination of the emerging trends, challenges, and prospective developments in the realm of omics research. As the field continues to evolve, overcoming challenges related to data integration, reproducibility, and standardization are underscored as crucial for the translation of these pioneering discoveries into improved patient care and the development of more effective, personalized therapeutic strategies. It is crucial to remember that the combination of omics tools and biotechnology will have significant effects on how medicine and healthcare are delivered in the future. As a result, it is essential to maintain research and development in this field to ensure that all future healthcare-related exigencies can be met with the most advanced and innovative solutions possible.

Keywords: Biotechnology, Drug interactions, Drug repurposing, High-throughput screening, Systems biology, Microbial biotechnology, Omics tools, Personalized medicine, Pharmacogenomics, Polypharmacology.

* **Corresponding author Israel Valencia Quiroz:** Phytochemistry Laboratory, UBIPRO, Superior Studies Faculty (FES)-Iztacala, National Autonomous University of Mexico (UNAM), Tlalnepantla de Baz, México State, 54090, México; Tel: +525572731888; E-mail: israelv@unam.mx

INTRODUCTION

The study of metabolites and metabolism, often known as metabolomics, occupies a distinctive position in the rapidly changing field of drug interaction research. Researchers can use this field as a conduit to investigate the complex intersections of gene-environment interactions. Notably, this particular focus stands in stark contrast to the emphasis on genes and genetic risk scores, which predominantly indicate potential future outcomes. Rather, metabolic profiling and phenotyping reflect ongoing processes in the present. In summary, metabolomics not only makes it easier to identify illness indicators such as endogenous metabolites (those produced by genes) and exogenous metabolites (those produced by environmental variables) but also offers previously unattainable insights into the underlying causes of diseases. With enhanced accessibility to metabolomics assays, these fresh perspectives are catalyzing a seismic shift in the way drugs are discovered, developed, delivered, and dosed. It is clear that this paradigm shift has the potential to transform the process of finding new drugs and cause a major shift in the medical industry [1].

In the realm of scientific research, a considerable obstacle that arises pertains to the unification of diverse omics data sets, a crucial procedure that facilitates the generation of more knowledgeable decisions about therapeutics. Despite the intricate nature of this process, it has become a rather commonplace occurrence to incorporate and blend exome and RNA-seq data, namely, through the fusion of genomics and transcriptomics, particularly when examining a tumor, as a means of unearthing profound insights into prospective therapeutic targets that can be influenced by drugs. This process, which is referred to as pharmacogenomics, is progressively gaining traction [2].

Pharmacogenomics is the field of study characterized by the discerning application of genomics data and other "omics" information for the purpose of guiding, informing, and individualizing drug therapy. A long time has passed since the initial introduction of the concept of pharmacogenetics by Arno Motulsky, and the progressions that have been achieved in this realm since then are noteworthy. It is undeniable that genetics has a significant impact on therapeutic efficacy and the likelihood of unexpected drug reactions and that genomics data can be used to enhance efficacy and reduce adverse reactions [2].

While the field of transcriptomics has been successful in elucidating gene expression modifications following drug exposure, proteomics has enabled researchers to delve into the intricate interactions between drugs and targets and the subsequent impacts on protein expression. However, the study of metabolomics involves a thorough characterization of metabolites and metabolism in biological systems. This field of "omics" science is quickly expanding. Numerous new biomedical applications have been made possible by recent developments in metabolomics technology. In particular, the study of metabolomics is beneficial for identifying novel drug targets, detecting diseases, understanding disease mechanisms, determining individual treatment plans, and precisely tracking therapeutic outcomes [1].

USING HIGH-THROUGHPUT SCREENING TO FIND NEW DRUGS

With the emergence of high-throughput technologies, there has been a proliferation of genome-scale datasets that have come to be known as multiomics data. The landscape of drug discovery research has undergone considerable upheaval as a result of this advancement, which has had a profound impact on the discipline. This vast collection of data comprises different types of data, such as genome sequencing data, which is known as genomics. Additionally, there are genome-wide RNA-sequencing data, which are referred to as transcriptomics. Furthermore, there are methylation and histone modification data, which fall under the category of epigenomics. Last, there is mass spectrometry protein data, which is known as proteomics [3].

Over the past four decades, the prevailing pattern for the exploration and establishment of drugs has typically involved an array of intricate stages. The procedure starts with the identification and determination of disease-triggering genes using technologies such as whole-genome sequencing, genome-wide association studies (GWAS), or pedigree analysis. The next step in development is the cloning of the discovered genes. High-throughput screening is performed on the purified target proteins to find potential therapeutic leads. These leads are then improved upon and evaluated in animal models, ultimately leading to human trials [1].

The field of metabolomics, despite being primarily focused on metabolites, has a multitude of applications that extend across a vast array of domains. Human and animal health, the identification of biomarkers, the development of new drugs, plant biology, microbiology, food chemistry, and environmental monitoring are just a few of these domains. Metabolomics' versatility in being able to evaluate a wide variety of substrates is the reason for its broad applicability. These substrates include liquids such as water, effluent, and biofluids as well as solids such as tissues, soil, and biological waste [1].

Genomics and transcriptomics approaches have unquestionably played a crucial part in the identification of prospective therapeutic targets and biomarkers that are suggestive of treatment response. The multiomics approach has been widely employed to supplement these methodologies [1]. Furthermore, proteomics and metabolomics tools, which are scientific techniques used to study the proteome and metabolome of an organism, respectively, have demonstrated their significant efficacy in the process of screening compound libraries for innovative inhibitors and activators. This has greatly contributed to the multipronged approach that is currently employed in modern drug discovery, which involves the use of various technologies and methodologies to identify potential drug candidates [1].

Personalized Medicine and Pharmacogenomics

The field of personalized medicine, which is experiencing vigorous growth, has seen significant progress due to swift advancements in high-throughput technologies and systems approaches. The traditional healthcare model, which historically concentrated only on the diagnosis and treatment of diseases, is currently being transformed by this dynamic evolution into a more all-encompassing strategy that includes predictive and preventative medicine as well as personalized health monitoring. Furthermore, the combination of individual genomics data with ongoing, global molecular component monitoring is projected to have a tremendously positive impact on individualized health treatment and provide real-time insights into physiological conditions [4].

Personalized medicine has entered a fascinating new age with the introduction of cutting-edge genetic tools such as CRISPR–Cas, particularly in the context of treatments for viral infections. This technique enables researchers to precisely change particular DNA sequences in living things. Numerous prospects for personalized medicine have emerged as a result of its accuracy, cost, and versatility. For instance, these developments may have a considerable impact on the treatment of viral infections such as HIV or hepatitis. Here, the genome of the patient might be examined to find particular genetic characteristics that affect their receptivity to the virus or reaction to treatment (Fig. **1**).

It is obvious that the CRISPR–Cas technology's enormous potential for personalized medicine can change the antiviral therapy industry and open the door for future therapeutics that are more specifically targeted.

At the forefront of the current transformation in the field of medicine lies a novel methodology that has garnered significant attention and interest, known as pharmacometabonomics. Predicting pharmacological effects using predose biofluid metabolite profile analysis is a novel and cutting-edge method that accounts for the intricate interplay between genetic and environmental factors that affect human physiology. This sophisticated and powerful capability empowers clinicians with valuable foresight, enabling them to make informed decisions and determine in advance which drugs will be efficacious and safe for their patients. As a critical and indispensable pillar of personalized medicine, pharmacometabonomics holds tremendous potential for improving therapeutic outcomes, specifically by aiding the selection of drugs that are both effective and safe for individual patients, thereby maximizing the benefits and minimizing the risks associated with medical treatment [5]. The methodology of implementing personalized medicine represents a revolutionary approach toward the future of medical treatment, wherein therapeutic regimens are customized and personalized in accordance with the unique biological characteristics and profiles of individual patients, thus ultimately augmenting the efficiency, effectiveness, and success of the treatment while simultaneously diminishing the probability and likelihood of potential adverse reactions.

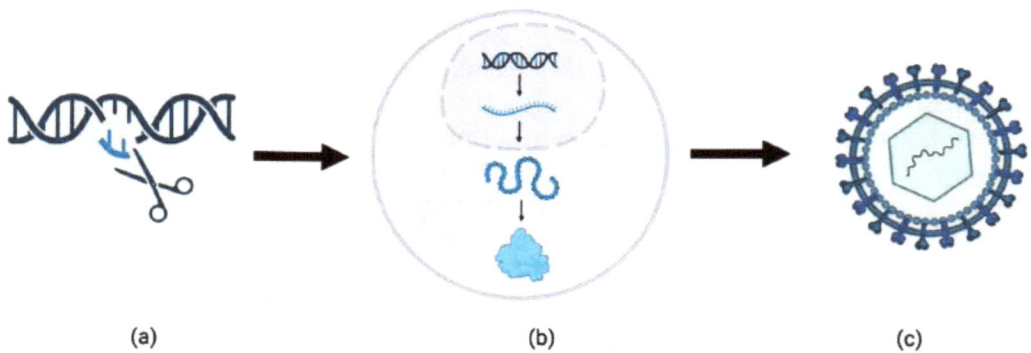

       (a)   (b)   (c)

Fig. (1). The Facets of CRISPR–Cas Technology in Personalized Antiviral Therapy: In this figure, we aim to elucidate the multifarious aspects of CRISPR–Cas technology in personalized medicine for viral disease treatment. First, panel (**a**) symbolically represents the CRISPR–Cas system as molecular scissors capable of making precise cuts in specific DNA sequences with utmost accuracy. This capacity allows for the targeted and specific modification of genetic information, thereby paving the way for a revolution in the field of genetic engineering. Moving on to panel (**b**), we see a cell expressing a gene of interest as a protein. After the CRISPR–Cas-mediated gene editing process, the modified gene can lead to the production of an altered protein with enhanced functions, such as improved immune recognition of virus-infected cells. This can be a game-changer in personalized antiviral therapies. Finally, panel (**c**) showcases a virus, representing the target of such personalized antiviral therapies. By leveraging the precision of CRISPR–Cas gene editing, patient-specific strategies can be developed to either directly target viral DNA or bolster the patient's own cellular defenses against the virus.

Drug-drug and Drug-metabolite Interactions

Drug-drug interactions (DDIs) are an essential component of pharmaceutical research and clinical practice that necessitates significant consideration. These interactions occur as a result of modifications in the plasma concentrations of one drug, commonly referred to as the 'victim', due to the introduction of another drug, causing the metabolism or transporter-mediated disposition of the "victim" to be inhibited or induced. These interactions may have severe consequences, with the 'victim' drug's exposure potentially being amplified or diminished by more than tenfold, resulting in potentially life-threatening outcomes. To assess the pharmacokinetic DDI risks of medications, a combined strategy that incorporates modeling methodologies and clinical investigations has emerged as the most cutting-edge method for improving the predictability and modeling of DDIs over time [6].

Drug-metabolite interactions (DMIs) essentially occur when a patient is administered or exposed to another drug that changes how they react to the original medication. The cytochrome P450 (CYP) enzyme, more specifically CYP3A4, which is involved in the metabolism of almost all tyrosine kinase inhibitors (TKIs), is one of the most significant participants in this complex process. As a result, there is a significant chance that TKIs and other medications that affect this complex metabolic route will interact. It is crucial to remember that cancer patients are more susceptible to DMIs since they frequently take several drugs, either for supportive care or toxicity management. Due to the broad therapeutic window of most common medications, the effects of DMIs are typically insignificant; nevertheless, in case of anticancer therapies, even small changes in drug metabolism and pharmacokinetics can have severe clinical effects [7].

It is essential to pay close attention to drug-metabolite interactions in addition to the complexity of drug-drug interactions. These types of interactions entail intricate interrelationships between a pharmacological agent and its ensuing metabolites, which are generated as a consequential consequence of the process of drug metabolism [8]. To explicate in a straightforward manner, the concept of drug-metabolite interactions pertains to the alteration of a drug's metabolic pathway, whereas drug-drug interactions refer to modifications in the drug's effects as a result of the introduction or coadministration of another pharmacological agent. It should not go unnoticed that both forms of interactions wield considerable influence over the effectiveness and safety of pharmaceutical interventions.

The utilization of omics instruments in elucidating the molecular mechanisms that underlie drug-drug and drug-metabolite interactions presents a highly auspicious avenue of inquiry. Omics technologies are positioned to offer insights into the complex interplay between drugs, their metabolites, and the proteins and pathways that they impact. For example, proteomics and metabolomics analyses can facilitate the detection of prospective drug-drug interactions and elucidate the ramifications of such interactions on cellular processes [9].

Drug Repurposing and Polypharmacology

The employment of omics tools has surfaced as a highly valuable resource in the discernment of novel therapeutic uses for extant drugs, as well as in the explication of the intricate interplay of drugs with manifold targets. The act of drug repurposing, wherein preexisting drugs are designated for new therapeutic applications, has seen significantly progress due to the implementation of omics technologies. Through meticulous examination of the influence of drugs on gene expression, protein levels, and metabolic pathways, researchers can uncover fresh targets and indications for previously sanctioned drugs. In addition, omics tools play a pivotal role in illuminating the multifarious interconnections between drugs and their targets, a notion recognized as polypharmacology.

Omics technologies encompassing genomics, proteomics, and metabolomics have proven to be highly effective in identifying fresh targets for the repurposing of drugs. Researchers can unveil new therapeutic applications for existing drugs by delving into the intricate molecular profiles of diseases and drugs. This approach is particularly beneficial in the case of rare diseases or illnesses that are limited in treatment options. For example, a drug designed initially for the treatment of cancer may be repurposed to tackle a rare genetic disorder based on the distinct molecular makeup of the drug [10].

Despite the positive outlook for the future, it is clear that there is a pressing requirement for the development of more extensive platforms dedicated to the analysis of data. Although the benefits that can be derived from big data are undeniable, particularly in terms of the identification of opportunities for repurposing, there continue to be significant obstacles that must be overcome in relation to data access and integration, specifically concerning clinical data such as clinician notes that are contained in patient case records. Consequently, it is imperative that advanced technological solutions be implemented to minimize the need for manual curation and to facilitate the integration of a wide range of omics data types. Such innovative practices would enable subsequent analyses to be conducted in a more sophisticated manner and the results to be presented in a format that is accessible to individuals who are not experts in the field [11].

A new and cutting-edge technique for drug discovery called polypharmacology promotes the creation of extremely potent medications by simultaneously modulating several targets. The normal drawbacks related to the use of single-target medications or the combination of several drugs may be lessened by this unique paradigm, which presents a novel pathway for the realization of enhanced pharmacological efficacy. In essence, the concept of polypharmacology signifies the utilization of a single drug to engage and modulate multiple biological targets, thereby opening up new vistas in drug discovery and development [10].

Systems Biology and Network Analysis

The integration discussed in this chapter serves as an example of how individualistic genomics data are combined with long-term global surveillance of molecular elements that reflect current physiological conditions. The advancement of high-throughput technologies has made this union possible. These aforementioned high-throughput sequencing and mass spectrometry-based technologies give researchers and medical professionals the unmatched ability to closely examine genomes, transcriptomes, proteomes, metabolomes, and other "omics" data. A comprehensive examination of health and disease is made possible by the convergence of "omics" data, opening up new possibilities for preventative medicine and individualized health monitoring [12].

One of the most noteworthy methods for comprehending human disease is the field of network medicine, which involves a comprehensive examination of the intricate network of interactions that occur among various cellular components, including but not limited to genes, proteins, and metabolites. This particular strategy is predicated on the fundamental insight that an anomaly in a single effector gene product is infrequently, if ever, the cause of a disease phenotype. Instead, it results from numerous pathobiological processes that communicate with one another within a sophisticated and intricate network of intricate components [13].

Furthermore, it is worth noting that network analysis tools are an invaluable resource regarding the visualization and analysis of the intricate interactions that occur between various drugs, their respective targets, and the numerous cellular pathways that they influence. Such analytical techniques enable researchers to develop a far better grasp of the underlying systems that control drug interactions, offering a more thorough comprehension of their effects [14]. As we embark upon a continued expedition into the vast and intricate realm of systems biology and network analysis, it is highly probable that the utilization of these cutting-edge tools will undeniably serve as a pivotal factor in molding and refining our comprehension and regulation of the multifaceted and dynamic field of health and disease.

Microbial Biotechnology and Drug Interactions

Within this chapter, we shall undertake a deep dive into the intricacies of the employment of the tools associated with omics in the pursuit of furthering our understanding of the interactions of drugs as they pertain to microbial systems. These interactions are inclusive of the mechanisms related to antibiotic resistance as well as the function of the microbiome in terms of mediating drug metabolism.

The emergence of omics technologies has been a significant development in the realm of microbial systems research. These innovative tools have proven to be invaluable resources in studying drug interactions and their effects. One area in which omics technologies have been particularly noteworthy is in their application toward exploring the development of antibiotic resistance. Through the utilization of the Parallel Annotation and Reassembly of Functional Metagenomic Selections (PARFuMS) method, researchers have been able to identify antibiotic resistance genes in nonpathogenic soil-dwelling bacteria. Interestingly, it has been discovered that these genes share exact nucleotide identity with an abundance of diverse human pathogens. Also discovered by the research is that mobile DNA elements surround several resistance genes that are colocalized within long stretches of perfect nucleotide similarity. This shows that multidrug resistance cassettes have recently been horizontally transferred from soil to the clinic, and it also draws attention to the possibility of conjugation or transformation as horizontal gene transfer routes [15].

Furthermore, it should be mentioned that the examination of the role of the human microbiome in medication metabolism and response has proven to be quite beneficial when using omics methods. The chemical modification of xenobiotics by the human gut microbiota is a topic of particular importance in this field of research. It is crucial to recognize that the examples of xenobiotic metabolism that have been investigated thus far probably only account for a small portion of the changes taking place within the gut ecosystem. This emphasizes the critical function of metabolomics in human patients, as it enables the identification of hitherto unknown gut bacteria metabolic pathways. When evaluating transformations that rely on extra microbial and host activities or interactions, it is crucial to take these processes into account. Therefore, it is essential for the sensible application of functional meals or prebiotics in the treatment of illnesses such as metabolic disease and malnutrition to have a thorough molecular understanding of how gut microorganisms digest dietary components [16].

As we progress further in the utilization of omics tools in the vast domain of microbial biotechnology, we are inevitably positioned to amass an even greater multitude of in-depth and comprehensive understandings regarding the intricate and multifaceted interdependence of microorganisms, pharmaceuticals, and overall human well-being.

ETHICAL, LEGAL, AND SOCIAL IMPLICATIONS (ELSI) OF OMICS TECHNOLOGIES IN BIOTECHNOLOGY

The chapter at hand embarks upon a rigorous examination of the considerable ethical predicaments deeply ingrained within the grandiose endeavors to demarcate the genetic foundations of human physiology, pathology, and drug response utilizing omics technologies within the domain of biotechnology.

Foremost among these ethical dilemmas is the possibility of genetic data procured through the misuse of research. If this information were to be handled improperly, it could potentially cause harm to the participants of the study or even to individuals whose genotype was being scrutinized but who are not part of the research. For instance, it is conceivable that these data could be divulged to other entities, such as law enforcement agencies or insurance companies, which raises a profound concern regarding the likelihood of misuse [17].

Despite the concerns highlighted above, the Human Genome Project (HGP) is an example of how such endeavors might have significant positive effects. By creating a sequence of the human genome, sequencing model organisms, developing high-throughput sequencing technologies, and, most importantly, thinking critically about the moral and social conundrums these technologies raise, the HGP has greatly benefited the fields of biology and medicine. By utilizing economies of scale and the coordinated efforts of a global team, the HGP was able to accomplish its goal much more quickly than if the genome had been sequenced gene-by-gene in isolated labs. Importantly, government backing for the HGP was motivated by the possibility of economic gains [18].

Clinical sequencing laboratories are strongly encouraged to actively pursue and meticulously report highly specific classes or types of mutations that have been thoroughly outlined and expounded upon during their extensive research endeavors. It is of utmost importance to place critical emphasis on properly and promptly notifying the patient about any and all potential incidental findings that may arise during pretest patient discussions, the actual clinical testing process, and the subsequent reporting of the obtained results [19].

In exploring the complex landscape of genomics medicine, an advanced and rapidly progressing field, we begin by unraveling numerous ethical considerations. The first ethical concern that merits our attention is genetic discrimination, which refers to the possibility of discrimination by employers and insurance companies based on an individual's genetic data. This issue presents a significant ethical quandary in genomics medicine, as it has the potential to cause harm to individuals and undermine their autonomy.

The second ethical concern that needs to be addressed is informed consent, which is essential in genomics medicine, given that patients' comprehension of the implications, potential hazards, and advantages of genetic testing may be limited. Healthcare practitioners should give patients clear and understandable information about the consequences and potential results of genetic testing to ensure that they are properly informed.

The third ethical concern that we need to address is stigmatization, which refers to the potential for individuals to face stigmatization based on their genetic information. Stigmatization can lead to negative consequences for individuals, including discrimination, social isolation, and psychological distress. To mitigate the risk of stigmatization, healthcare professionals and policymakers should collaborate to develop guidelines and policies that promote nondiscrimination and social inclusion.

Other issues include equity and access, which pertain to the possibility of unequal access to genetic testing and genomics therapy. Unfair access to genomics medicine and genetic testing has the potential to widen existing health inequalities. Therefore, regardless of socioeconomic level or other demographic characteristics, healthcare professionals and governments should cooperate to guarantee that everyone has equitable access to genomics medicine and genetic testing and can enjoy the advantages of such technologies [20].

The above section, therefore, serves to highlight and emphasize the intricate and multifaceted ethical, legal, and social aspects that necessitate careful and deliberate contemplation as we persist in the utilization and progression of omics technologies within the realm of biotechnology.

FUTURE PERSPECTIVES AND CHALLENGES

The overarching theme under examination in this particular discourse pertains to the investigation of nascent trends and cutting-edge technologies that are presently being developed and utilized in the field of omics research. The primary focus of this inquiry revolves around the seemingly boundless potential that these innovations have to fundamentally alter and redefine the scope and nature of studies pertaining to drug interactions, as well as the broader landscape of biotechnology at large. Coextensive with this line of inquiry, it is also essential to undertake a thorough exploration of the various challenges and limitations that researchers and practitioners alike inevitably confront throughout this exhilarating and dynamic journey of discovery.

The incessant development of omics technologies, including but not limited to single-cell sequencing and refined mass spectrometry techniques, presents an unparalleled opportunity to intensify our comprehension of drug interactions at a much more intricate level of detail. High-resolution structural techniques, for example, play an indispensable role in steering the optimization of bioactive substances to transform them from low-efficiency hits into promising leads. These methods make it easier to reveal the target protein's three-dimensional structure and how it interacts with small molecule fragments. Following this, the acquired information serves as a prototype for enhancing the structure of these fragments, thereby amplifying their binding affinity and selectivity toward the target protein. With these conspicuous improvements, the fragments can be advanced into leads and potential drug candidates for ensuing development [21].

It is projected that the convergence and integration of omics data with artificial intelligence and machine learning would considerably improve the accuracy and efficacy of drug interaction prognostications. The development and expansion of machine learning, especially deep learning, along with the widespread use of modern computer hardware such as graphical processing units (GPUs), have unlocked a variety of applications inside the realm of pharmaceutical firms. These machine-learning applications span a wide range of tasks and endeavors, including the validation of targets, identification and recognition of prognostic biomarkers, analysis and scrutiny of digital pathology data in clinical trials, and prognosticating and predicting drug toxicity [22].

Concurrently and in a concomitant manner, the scientific technique of single-cell sequencing is forecasted to furnish a highly detailed comprehension of the intricate and multifarious nature of cellular heterogeneity, as well as the fundamental and latent mechanisms of drug interactions at the level of the individual cell, thereby providing an unprecedented level of insight and depth into the workings of the biological system [23].

Despite the remarkable progress made in this field, numerous formidable challenges continue to persist. The standardization of data formats, seamless integration of multiomics data, and translation of omics findings into clinically actionable insights are still considered significant roadblocks in this area of research. The integration of data, for instance, remains a critical challenge owing to the inherent complexities and computational intensity involved in amalgamating and analyzing vast, diverse data sets. Therefore, it is necessary to overcome these challenges to achieve the desired outcomes in this field [3]. Furthermore, it is imperative to acknowledge and address the pressing necessity of guaranteeing the replication of data and implementation of uniform procedures across various platforms and laboratories. This critical concern demands a carefully crafted and deliberate approach to ensure its successful resolution.

Omics technologies have significant ethical, legal, and social ramifications that should not be understated. As we continue to make progress in this field, it is crucial that we carefully examine issues related to data privacy, informed consent, and the responsible management of incidental findings in clinical settings. It is essential to ensure that the use of omics technologies is conducted responsibly, with a focus on safeguarding patient autonomy and privacy at every stage of the process. This central theme serves to highlight these critical considerations and promote a balanced and responsible approach to the future of omics research and its applications in the field of biotechnology.

CONCLUSION

In conclusion, it is conspicuously apparent that the collaborative interdependence of omics tools and biotechnology has paved the way for extraordinary advancements in our comprehension of pharmaceutical interactions at the molecular stratum. By harnessing the amalgamation of myriad omics technologies and computational methodologies, scholars are unraveling the convoluted mechanisms that underlie drug interactions with an unparalleled degree of comprehensiveness and exactitude that was hitherto unachievable.

Pharmacology and drug repurposing are two fields that have come together, catalyzing the development of new and creative methods. These methods have in turn sparked a paradigm shift in favor of the use of more efficient and individualized therapy techniques. A deeper knowledge of the function played by microorganisms in drug metabolism and resistance has also been made possible by major and notable breakthroughs in omics technologies and their application to microbial systems. Therefore, these discoveries have created brand-new channels for intervention.

Recently, there has been a concomitant rise in the use of high-throughput technologies that have facilitated the comprehensive profiling of health and disease globally, thereby paving the way for the realization of more individualized health monitoring practices and preventative medicine approaches. Moreover, the convergence of omics data analysis with artificial intelligence and machine learning has exhibited significant promise by virtue of its potential to augment the precision and effectiveness of drug interaction predictions.

The present terrain of dynamic, cross-functional interaction highlights the potential of a more subtle comprehension of biological structures and the establishment of more potent therapeutic interventions. With the steady progression of these technologies, the amalgamation of emerging patterns such as single-cell sequencing, high-resolution structural techniques, and advanced artificial intelligence algorithms can completely alter the landscape, thereby elevating drug interaction studies to a more refined level and augmenting the effectiveness of biotechnology.

The advent of an auspicious future, which holds much promise, has its own set of difficulties. Among these challenges, the assimilation of incongruous omics data categories, the requirement for uniform data structures, and the conversion of omics discoveries into clinically viable perceptions continue to pose significant hindrances. The remediation of these obstacles is an indispensable measure to harness the complete potential of omics technologies and ensure that these advancements translate into tangible enhancements in patient care.

As we continue to traverse the uncharted territory of omics research, we must acknowledge the salient ethical, legal, and social implications that cannot be overlooked. To guarantee the utmost integrity and uprightness regarding the handling of data, it is imperative that we account for the preservation of patient privacy, the acquisition of informed consent, and the potential for genetic discrimination. These are all crucial considerations that must be accounted for to ensure the responsible application of omics technologies.

The amalgamation of various omics tools and biotechnology presents a significant and substantial opportunity to redefine and reshape our comprehension of drug interactions and the all-encompassing realm of biotechnology. The expedition ahead is replete with prospects and trials, and the perpetual advancement and progression of these cutting-edge technologies will serve as a crucial and pivotal factor in shaping and forging the future of personalized medicine.

REFERENCES

[1]   Wishart DS. Emerging applications of metabolomics in drug discovery and precision medicine. Nat Rev Drug Discov 2016; 15(7): 473-84.
      [http://dx.doi.org/10.1038/nrd.2016.32] [PMID: 26965202]

[2]   Weinshilboum RM, Wang L. Pharmacogenomics: Precision medicine and drug response. Mayo Clin Proc 2017; 92(11): 1711-22.
      [http://dx.doi.org/10.1016/j.mayocp.2017.09.001] [PMID: 29101939]

[3]   Ritchie MD, Holzinger ER, Li R, Pendergrass SA, Kim D. Methods of integrating data to uncover genotype phenotype interactions. Nat Rev Genet 2015; 16(2): 85-97.
      [http://dx.doi.org/10.1038/nrg3868] [PMID: 25582081]

[4]   Chen R, Snyder M. Systems biology: Personalized medicine for the future? Curr Opin Pharmacol 2012; 12(5): 623-8.
      [http://dx.doi.org/10.1016/j.coph.2012.07.011] [PMID: 22858243]

[5]   Everett JR. Pharmacometabonomics in humans: A new tool for personalized medicine. Pharmacogenomics 2015; 16(7): 737-54.
      [http://dx.doi.org/10.2217/pgs.15.20] [PMID: 25929853]

[6]   Tornio A, Filppula AM, Niemi M, Backman JT. Clinical studies on drug–drug interactions involving metabolism and transport: Methodology, pitfalls, and interpretation. Clin Pharmacol Ther 2019; 105(6): 1345-61.
      [http://dx.doi.org/10.1002/cpt.1435] [PMID: 30916389]

[7]   Teo YL, Ho HK, Chan A. Metabolism-related pharmacokinetic drug–drug interactions with tyrosine kinase inhibitors: Current understanding, challenges and recommendations. Br J Clin Pharmacol 2015;

79(2): 241-53.
[http://dx.doi.org/10.1111/bcp.12496] [PMID: 25125025]

[8]   VandenBrink BM, Isoherranen N. The role of metabolites in predicting drug-drug interactions: focus on irreversible cytochrome P450 inhibition. Curr Opin Drug Discov Devel 2010; 13(1): 66-77.
[PMID: 20047147]

[9]   Yıldırım MA, Goh KI, Cusick ME, Barabási AL, Vidal M. Drug—target network. Nat Biotechnol 2007; 25(10): 1119-26.
[http://dx.doi.org/10.1038/nbt1338] [PMID: 17921997]

[10]   Anighoro MA, Rgen Bajorath J, Rastelli G. Polypharmacology: Challenges and opportunities in drug discovery. J Med Chem 2014; 57(19): 7874-87.
[http://dx.doi.org/10.1021/jm5006463]

[11]   Pushpakom S, Iorio F, Eyers PA, *et al.* Drug repurposing: Progress, challenges and recommendations. Nat Rev Drug Discov 2019; 18(1): 41-58.
[http://dx.doi.org/10.1038/nrd.2018.168] [PMID: 30310233]

[12]   Weston AD, Hood L. Systems biology, proteomics, and the future of health care: toward predictive, preventative, and personalized medicine. J Proteome Res 2004; 3(2): 179-96.
[http://dx.doi.org/[https://doi.org/10.1021/pr0499693]] [PMID: [PMID: 15113093]]

[13]   Barabási AL, Gulbahce N, Loscalzo J. Network medicine: A network-based approach to human disease. Nat Rev Genet 2011; 12(1): 56-68.
[http://dx.doi.org/10.1038/nrg2918] [PMID: 21164525]

[14]   Dudley JT, Deshpande T, Butte AJ. Exploiting drug-disease relationships for computational drug repositioning. Brief Bioinform 2011; 12(4): 303-11.
[http://dx.doi.org/10.1093/bib/bbr013] [PMID: 21690101]

[15]   Forsberg KJ, Reyes A, Wang B, *et al.* The shared antibiotic resistome of soil bacteria and human pathogens. Science 2012; 337(6098): 1107-11.
[http://dx.doi.org/https://doi.org/10.1126%2Fscience.1220761] [PMID: 22936781]

[16]   Koppel N, Rekdal VM, Balskus EP. Chemical transformation of xenobiotics by the human gut microbiota. Science 2017; 356(6344): eaag2770.
[http://dx.doi.org/10.1126/science.aag2770] [PMID: 28642381]

[17]   Roden DM, George AL Jr. The genetic basis of variability in drug responses. Nat Rev Drug Discov 2002; 1(1): 37-44.
[http://dx.doi.org/10.1038/nrd705] [PMID: 12119608]

[18]   Hood L, Rowen L. The human genome project: Big science transforms biology and medicine. Genome Med 2013; 5(9): 79.
[http://dx.doi.org/10.1186/gm483] [PMID: 24040834]

[19]   Green RC, Berg JS, Grody WW, *et al.* ACMG recommendations for reporting of incidental findings in clinical exome and genome sequencing. Genet Med 2013; 15(7): 565-74.
[http://dx.doi.org/10.1038/gim.2013.73] [PMID: 23788249]

[20]   Vassy JL, Korf BR, Green RC. How to know when physicians are ready for genomic medicine. Sci Transl Med 2015; 7(287): 287fs19.
[http://dx.doi.org/10.1126/scitranslmed.aaa2401] [PMID: 25971999]

[21]   Macarron R, Banks MN, Bojanic D, *et al.* Impact of high throughput screening in biomedical research. Nat Rev Drug Discov 2011; 10(3): 188-95.
[http://dx.doi.org/10.1038/nrd3368] [PMID: 21358738]

[22]   Vamathevan J, Clark D, Czodrowski P, *et al.* Applications of machine learning in drug discovery and development. Nat Rev Drug Discov 2019; 18(6): 463-77.
[http://dx.doi.org/10.1038/s41573-019-0024-5] [PMID: 30976107]

[23]   Macaulay IC, Voet T. Single cell genomics: Advances and future perspectives. PLoS Genet 2014; 10(1): e1004126.
[http://dx.doi.org/10.1371/journal.pgen.1004126] [PMID: 24497842]

CHAPTER 10

Sharing is Caring: Drug Repurposing among Leading Diseases

Verónica García-Castillo[1], Eduardo López-Urrutia[1], Carlos Pérez-Plasencia[1,2] and Adriana Montserrat Espinosa-González[3,*]

[1] *Genomics Lab, Biomedicine Unit, FES-Iztacala, National Autonomous University of Mexico, Tlalnepantla, 54090, Mexico*

[2] *Genomics Lab, National Cancer Institute (INCan), Tlalpan, Mexico City, 14080, Mexico*

[3] *Phytochemistry Laboratory, UBIPRO, Superior Studies Faculty (FES)-Iztacala, National Autonomous University of Mexico (UNAM), Tlalnepantla de Baz, México State, 54090, México*

Abstract: The process of drug development is time-consuming and resource-intensive, but drug repurposing offers an alternative by using already approved drugs to treat different diseases. Drug repurposing candidates can be identified through computational and experimental approaches, which are often combined. Traditionally, drug repurposing is considered when developing a custom drug is not feasible, but recent findings regarding the cross-talk between cellular mechanisms and pathways that are altered among disease states suggest that multipurpose drugs may be the key to simultaneously treating multiple diseases. This chapter reviews published reports on drug repurposing for five of the most threatening diseases to human health today: Alzheimer's disease, arthritis, diabetes mellitus, cancer, and COVID-19, highlighting promising candidates, challenges, and potential future directions for research.

Keywords: Alzheimer's disease, Arthritis, Cancer, COVID-19, Diabetes mellitus, Drug development, Drug repurposing.

INTRODUCTION

Drug development is a complex process that can take years to complete. After its discovery, which alone can take several attempts spanning years, candidate drugs must undergo a series of trials to deem them safe, effective, and practical to use as a therapy. These tests can take close to twenty years and have a low probability of success. Drug repurposing, also known as drug repositioning, offers an alternative

* **Corresponding author Adriana Montserrat Espinosa-González:** Phytochemistry Laboratory, UBIPRO, Superior Studies Faculty (FES)-Iztacala, National Autonomous University of Mexico (UNAM), Tlalnepantla de Baz, México State, 54090, México; Tel: +525556231136; E-mail: adriana.espinosa@iztacala.unam.mx

Israel Valencia Quiroz (Ed.)

to this resource-intensive process. The rationale behind drug repurposing is to discover new uses for existing drugs that have already been tested to be safe, effectively bypassing at least the first set of tests and saving substantial amounts of time and resources [1]. This approach has succeeded in finding novel uses for several drugs. For instance, the PDE5 inhibitor sildenafil, initially intended as a treatment for systemic hypertension and later discovered to be an effective treatment for erectile dysfunction, has found new applications in breast and prostate cancer, among other cancer types [2]. Another successful, albeit noncanonical, example is thalidomide, a nonaddictive sedative marketed to pregnant women in the 1960s. After being infamously teratogenic, it was banned; however, it is now used as a treatment for various cancers and inflammatory skin disorders [3].

Drug repurposing is a relatively recent concept and has only been documented since the mid-1980s. Since then, approaches to this potentially beneficial concept have moved from serendipitous to systematic. Today, drug repurposing strategies are broadly classified into computational and experimental approaches. Computational strategies use big data analytics to find repurposing candidates through their association with gene or pathway regulation, large-scale drug screen results, and molecular docking. Experimental approaches encompass binding assays such as affinity chromatography and *in vitro* experiments to evaluate the effectiveness of a repurposing candidate [4]. Repurposed drugs are great alternatives to treat rare diseases where no licensed treatment is available, and the resource-intensive process of developing a custom drug is unfeasible [5].

Nevertheless, as the community searching for repurposing candidates and applications widens and the amount of available information increases, it has become evident that drugs can be repurposed even among the leading, most studied diseases, exploiting the crosstalk between the cellular mechanisms and pathways altered in them. This crosstalk indicates that a few already-approved drugs may be able to simultaneously treat several diseases. In this chapter, we reviewed published information about drug repurposing for five of the most threatening diseases to human health today: Alzheimer's disease [6], arthritis [7], diabetes mellitus [8], cancer [9], and the recently emerged COVID-19 [10].

THE PANDEMIC IN THE ROOM

Although the strict lockdown measures have been lifted, the world is still dealing with the pandemic caused by the SARS-CoV-2 virus, the etiological agent of COVID-19. Several groups rushed to find treatment strategies for the novel disease [11], and as the average number of severe cases dwindles, concerns about its sequelae have become increasingly important [12, 13].

Due to its relative but substantial reduction in processing time, drug repurposing excels as a search strategy for COVID-19 treatments. For instance, Mirabelli *et al.* [14] found that amiodarone, ipratropium bromide, lactoferrin, lomitapide, remdesivir, and Z-FA-FMK, out of a library of over a thousand US Food and Drug Administration (FDA)-approved compounds, inhibited SARS-CoV-2 infection in a panel of cell lines. Of these compounds, lactoferrin stood out since it blocked viral attachment and enhanced the interferon response at concentrations as low as the nanomolar range.

Sonkar and colleagues [15] took a bioinformatic approach based on observed structural similarities between SARS-CoV-2 RNA-dependent RNA polymerase (RdRP) and familiar kinases such as JAK, ITK, PERK, and p38-MAPK. From a suite of 12 kinase inhibitors currently employed to treat gastric cancer, brepocitinib, decernoitinib, filgotinib, and ibrutinib bound SARS-CoV-2 RdRP with sufficient affinity to block its function, thus making them candidates for repurposing. A similar machine-learning approach suggested that baricitinib, an inhibitor of the AAK and JAK1/2 kinases, was effective against SARS-CoV-2 entry because AAK is involved in endocytosis. This mechanism was tested *in vitro* and in patients [16]. However, a 648-sample clinical study showed that treatment with the tyrosine kinase inhibitor imatinib did not necessarily lead to quicker COVID recovery. While strong evidence suggests that imatinib inhibits the Arg/Abl2 kinases, its administration was not associated with a better clinical outcome. The authors do not rule out the effectiveness of imatinib; instead, they argue that the metabolic variants within their sample may have affected the results [17].

A different bioinformatic approach, taken by Lucchetta and Pellegrini [18], was used to compare the gene expression patterns of patients with several conditions, including prostate and colorectal cancers and COVID-19. These authors developed DrugMerge, a software application capable of analyzing expression signatures and finding drugs that similarly affect them, ranking them according to their degree of similarity. Etoposide, a topoisomerase II inhibitor used to treat prostate cancer, was shown to affect a similar subset of genes in patients with COVID-19 infection. DrugMerge software is still in the process of being developed, and its developers are yet to incorporate considerations of pharmacological therapy, such as dosage, tissue specificity, and possible adverse side effects.

Given that COVID-19 can have broad systemic effects [19], several groups evaluated the repurposing of antimetabolic drugs for COVID-19 treatment. The results showed that several of these drugs can indeed aid in COVID-19 treatment.

For instance, methotrexate, an immunosuppressive drug that blocks JAK/STAT signaling, substantially attenuated the JAK1 and STAT3 phosphorylation mediated by the SARS-CoV-2-S1 subunit of the spike protein and increased IL6 production, leading to decreased ACE2 expression levels, and consequently, to reduced viral entry [20]. Ticlopidine, an antiplatelet drug that blocks purinergic receptors, has been proven to bind sigma-1 receptors (S1Rs) in the endoplasmic reticulum (ER), triggering the unfolded protein response (UPR) and thus leading to apoptosis. Coronavirus infections are highly dependent on the ER due to an exacerbated production of glycoproteins, so the augmented UPR caused by ticlopidine is substantially damaging not only SARS-CoV-2 but also SARS-CoV-1 and MERS-CoV, two other coronaviruses that have recently caused global-scale epidemics [21].

Cancer [22] and COVID-19 [23] both cause systemic inflammation, and dexamethasone is a potent anti-inflammatory drug that can treat both conditions. Dexamethasone has low solubility and can have severe side effects such as hypertension, hypoglycemia, and peptic ulcers, so current efforts for its repositioning are focused on more efficient and precise delivery systems. Dexamethasone-loaded nanoparticles have significantly improved disease-free survival in mice infected with SARS-CoV-2 [24].

Dantrolene is a ryanodine receptor antagonist used as a treatment for malignant hyperthermia with anti-inflammatory effects [25]. Wei and collaborators reported that dantrolene decreased sepsis or sepsis shock, which is caused by the strong, systemic inflammatory response elicited by the SARS-CoV-2 spike protein, in patients with acute COVID-19 [26].

An aspect of SARS-CoV-2 infection that is currently gaining increasing attention is the set of symptoms that linger after acute infection, regarded as a condition called long COVID. This syndrome may occur in all age groups and includes neuropsychiatric symptoms that might start during or after the infection. It is believed to arise from immune-related microvascular inflammation, hence the absence of viral markers [27]. Several reports have indicated an inverse correlation between antipsychotics and the likelihood of contracting COVID-19. In particular, chlorpromazine had anti-SARS-CoV-2 activity *in vitro*, although the exact effective dose is still to be determined. Fluvoxamine and sertraline blocked the sigma-1 receptor, preventing the SARS-CoV-2 spike protein from binding to it, thus blocking viral entry into the cell. Interestingly, the researchers who performed these studies did not test the effects of antipsychotic drugs on COVID-19; instead, they observed that the incidence and severity of long COVID-19 were lower in patients already subjected to neuropsychiatric treatment and later validated the findings [13]. Similarly, Kato and colleagues [28] found that clomipramine can also block SARS-CoV-2 entry by inhibiting clathrin-dependent endocytosis.

Finally, two groups of antidiabetic medications have been found to target the SARS-CoV-2 Mpro viral protease: DPP4 inhibitors and GPR 120 receptor agonists. Published data about these interactions mainly include silico docking studies, and these interactions still need to be experimentally verified. Gemigliptin, linagliptin, and evogliptin DPP4 inhibitors bind and decrease the activity of SARS-CoV-2 Mpro, suggesting that they might also be employed to treat COVID-19 [29]. Moreover, an analysis of 68 possible ligands with GPR 120 receptor agonist activity yielded eight with a score high enough to suggest therapeutic application. These compounds were functionally similar to linoleic acid, a free fatty acid that binds to the SARS-CoV-2 spike protein [30].

The silver lining of the COVID-19 pandemic is the rarely seen global cooperation in combating different aspects of the same problem. While vaccines may have been the key to controlling the emergence of new COVID cases, different treatment strategies have been developed just as rapidly. The above summary is far from exhaustive but offers a broad perspective of the strategies against COVID-19 developed in the realm of drug repositioning.

ANTI-DIABETICS: DRUGS TO RULE THEM ALL?

Isolated in the 1950s from the plant *Galega officinalis*, metformin is a first-line therapy for diabetes and is widely used as a monotherapy or in combination with other drugs. It is efficient at low doses, its maximum tolerated doses are reasonably high, and it produces relatively few side effects. Its popularity has risen in the last three decades since it was authorized in the USA [31]. Such popularity has led to increased attention to its pharmacokinetics, which has revealed that metformin participates in several cellular processes underpinning diverse conditions, including virtually all of those reviewed in this chapter: cancer, Alzheimer's disease, arthritis, and COVID-19 (Fig. **1**).

Metformin primary exerts an anticancer effect by suppressing the electron transport chain, which leads to cell cycle arrest through the AMPK and PI3K pathways and is upregulated in response to inhibition of the NF-κB pathway [32]. For cancer treatment, metformin is often administered together with chemotherapeutics such as everolimus in pancreatic and neuroendocrine tumors, gefitinib in non-small cell lung cancer, or paclitaxel in adenocarcinoma of the pancreas [33]. Metformin might not be the only plant-derived compound capable of modulating PI3K/Akt/mTOR signaling, as shown by several groups searching for phytochemicals with similar pleiotropic effects. Nurcahyanti *et al.* [34] reviewed as many as 717 compounds with possible antidiabetic activity, including alkaloids, flavonoids, polyphenols, triterpenoids, and steroids. From these, 43 modulated PI3K/Akt/mTOR signaling and thus had potential anticancer activity, opening new repurposing possibilities.

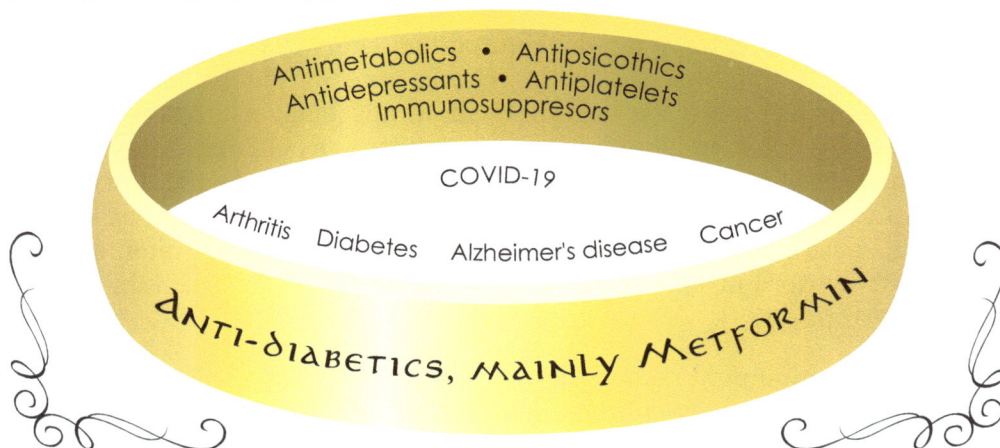

Fig. (1). Metformin leads a growing fellowship of repurposed drugs, in the quest against the most threatening diseases to human health today.

In a large meta-analysis including prospective and retrospective studies, Triggle and colleagues [35] found that metformin exerts an overall protective effect by suppressing proinflammatory pathways and protecting mitochondria. Moreover, they found that metformin had a protective effect on neural stem cells. This evidence suggests that metformin exerts a pleiotropic effect, a concept further explored by Morale and collaborators [36], who found that metformin can hinder cancer growth through the modulation of cell metabolism. Metformin inhibited tumor growth by preventing thyroid cancer cells from switching to anaerobic glycolysis. Conversely, it also inhibited the growth of prostate cancer-derived cells by arresting the cell cycle in the G1 phase.

The anti-inflammatory and immunoregulatory effects suggest that metformin may be able to treat COVID-19. It turned out that metformin interacts with two enzymes essential for the development of diabetes, angiotensin-converting enzyme 2 (ACE2) and dipeptidyl-peptidase 4 (DPP4), which act as receptors for the SARS-CoV-2 virus. Metformin stabilized ACE2 in respiratory tract cells, reducing the rate of SARS-CoV-2 infection; it also decreased DPP4 activity, disrupting viral attachment and controlling inflammatory processes. Diabetic patients with COVID-19 benefited the most from this approach, but the data suggested that metformin can be used to treat COVID-19 patients without diabetes [37].

In another broad application of metformin, Matsuoka and colleagues [38] found that it suppressed osteoclastogenesis and the inflammatory response in a human synovial fibroblast line, thus suggesting its potential use to treat rheumatoid arthritis. Other groups have reported positive results when administering metformin in combination with different drugs. For example, together with CoQ10, metformin inhibits osteoclastogenesis, one of the leading causes of rheumatoid arthritis. Inhibiting mTOR with drugs such as rapamycin is an emerging therapy for rheumatoid arthritis, but these mTOR inhibitors can also destabilize mitochondria; However, metformin reduces this side effect, increasing the utility of these drugs. Another therapeutic alternative under recent exploration is cyclooxygenase (COX)-2 inhibition, and while research is ongoing, evidence suggests that metformin enhances the positive effect of COX-2 inhibitors on rheumatoid arthritis [39].

However, metformin might not yet be an all-purpose drug. A 9-year retrospective study showed that the dipeptidyl-peptidase 4 (DPP4) inhibitors alogliptin, linagliptin, saxagliptin, sitagliptin, and vildagliptin were more effective than metformin in preventing recurrence in colorectal cancer patients [40]. In the following years, there will undoubtedly be more profound insights into the activities of metformin within the cell, which will most likely increase its applications.

Other diabetes treatments also have repurposing potential. For instance, in a retrospective study, Hendriks *et al.* [41] found that sulfonylurea derivatives, used to increase insulin levels in type 2 diabetes, reduce cancer cell proliferation by blocking PI3K/AKT signaling. Although more studies are needed, available evidence indicated that patients with diabetes treated with gliclazide had a lower cancer risk than those treated with glibenclamide.

While the potential of metformin as a polyvalent drug has not yet been fully explored, other common antidiabetic drugs are gaining attention as potential repurposing drugs, and the growing numbers of closely monitored diabetes mellitus patients offer a wealth of data to identify even more candidates.

DO NOT FORGET TO REPURPOSE FOR ALZHEIMER'S DISEASE

The multifactorial nature of Alzheimer's disease makes its treatment substantially challenging. For decades, the only available drugs for treating Alzheimer's disease were acetylcholinesterase inhibitors, based on the so-called cholinergic hypothesis, which attributes cognitive decline to the malfunction of acetylcholine-containing neurons [42]. The ongoing development of novel therapies for Alzheimer's disease follows two main strategies: decreasing the formation of amyloid plaques and neurofibrillary tangles mainly by reducing oxidative damage or reducing neuroinflammation to hinder the progression of the disease in its early stages [43]. Multiple well-studied drugs that have anti-inflammatory properties may be drug repurposing candidates for the treatment of Alzheimer's disease.

Lapatinib is a kinase inhibitor currently used in cancer treatment that inhibits EGFR and HER-2 receptors, as well as GSK-3β, P38-MAPK, c-Jun, and ERK kinases, reducing inflammation. In a study in rats, lapatinib decreased the gene expression levels of the oxidative stress-producing enzyme NOX- in addition to exerting anti-inflammatory effects; these effects counteracted the learning and memory declines observed in the untreated rats [44]. The EGFR inhibitors afatinib and ibrutinib downregulated the expression of glial acidic fibrillary protein (GFAP), cyclooxygenase-II, caspase-1, and nitric oxide synthase in cultured astrocytes and enhanced memory test results in mice. However, Mansour and colleagues discuss that the true challenge in repurposing EGFR inhibitors for Alzheimer's disease is to find compounds that clear the blood-brain barrier; lapatinib and ibrutinib seem promising in this regard [45]. Tofacitinib, a kinase inhibitor used for treating rheumatoid arthritis, seemed equally promising to Desai and collaborators [46], who performed a cohort study that included 4224 patients with rheumatoid arthritis treated with tofacitinib, searching for an association with decreased risk of developing Alzheimer's disease. The results were not as encouraging, as they found no difference in the risk of Alzheimer's disease in this cohort over the evaluated 10-year period. The contrast between these findings and the positive results obtained from an independent group [47] suggests that the timeframe of the Desai study may not have been optimal.

The anticancer effects of kinase inhibitors have been improved with other drugs, such as dantrolene [48], which is an example of drug repurposing in itself since it was developed as a muscle relaxant and quickly found to be effective as a treatment for malignant hyperthermia [25] and has been further repurposed for COVID-19 (see above), which also has repurposing potential for Alzheimer's disease. Shi and colleagues found that intranasal administration of dantrolene in a mouse AD model improved memory, both hippocampus-dependent and hippocampus-independent, in the early stages of the disease [49]. Moreover, dantrolene inhibited synapse decrease and promoted proliferation in a primary culture of pluripotent stem cells from Alzheimer's disease patients [50].

In addition to the obvious physical constraints of accessing the central nervous system, Alzheimer's disease's slow development poses substantial challenges to its treatment. However, the broad knowledge of the modulation of the inflammatory response gained from other study areas indicates that drug repurposing is a powerful tool for more efficient treatment strategies for this condition.

CONCLUSION

While this chapter is by no means a comprehensive record of all of the ongoing drug repurposing studies, it shows how popular drug repurposing is these days. Active research on the pathogenesis of diseases that have long affected humanity has generated vast amounts of information that is now possible to catalog and analyze. As a result, the scientific community can optimize resources and propose new treatments based on known drugs and their pharmacodynamics. We found the most dramatic instance of this situation is the high number of cancer drugs that offer potential applications in other diseases.

ACKNOWLEDGEMENTS

CONSEJO MEXIQUENSE DE CIENCIA Y TECNOLOGÍA. Fondo para la investigación científica y desarrollo tecnológico del Estado de México. Convocatoria: EDOMÉX-FICDTEM-2022-01. Financiamiento para investigación de mujeres científicas, COMECYT.

REFERENCES

[1]   Ashburn TT, Thor KB. Drug repositioning: Identifying and developing new uses for existing drugs. Nat Rev Drug Discov 2004; 3(8): 673-83.
[http://dx.doi.org/10.1038/nrd1468] [PMID: 15286734]

[2]   Cruz-Burgos M, Losada-Garcia A, Cruz-Hernández CD, *et al.* New Approaches in Oncology for Repositioning Drugs: The Case of PDE5 Inhibitor Sildenafil. Front Oncol 2021; 11: 627229.
[http://dx.doi.org/10.3389/fonc.2021.627229] [PMID: 33718200]

[3]   Amare GG, Meharie BG, Belayneh YM. A drug repositioning success: The repositioned therapeutic applications and mechanisms of action of thalidomide. J Oncol Pharm Pract 2021; 27(3): 673-8.
[http://dx.doi.org/10.1177/1078155220975825] [PMID: 33249990]

[4]   Pushpakom S, Iorio F, Eyers PA, *et al.* Drug repurposing: Progress, challenges and recommendations. Nat Rev Drug Discov 2019; 18(1): 41-58.
[http://dx.doi.org/10.1038/nrd.2018.168] [PMID: 30310233]

[5]   Roessler HI, Knoers NVAM, van Haelst MM, van Haaften G. Drug repurposing for rare diseases. Trends Pharmacol Sci 2021; 42(4): 255-67.
[http://dx.doi.org/10.1016/j.tips.2021.01.003] [PMID: 33563480]

[6]   Li X, Feng X, Sun X, Hou N, Han F, Liu Y. Global, regional, and national burden of Alzheimer's disease and other dementias, 1990–2019. Front Aging Neurosci 2022; 14: 937486.
[http://dx.doi.org/10.3389/fnagi.2022.937486] [PMID: 36299608]

[7]   Finckh A, Gilbert B, Hodkinson B, *et al.* Global epidemiology of rheumatoid arthritis. Nat Rev Rheumatol 2022; 18(10): 591-602.
[PMID: 36068354]

[8]   Saeedi P, Petersohn I, Salpea P, Malanda B, Karuranga S, Unwin N, *et al.* Global and regional diabetes prevalence estimates for 2019 and projections for 2030 and 2045: Results from the International Diabetes Federation Diabetes Atlas, 9th edition. Diabetes Res Clin Pr 2019; 157: 107843.

[9]   Sung H, Ferlay J, Siegel RL, *et al.* Global cancer statistics 2020: GLOBOCAN estimates of incidence and mortality worldwide for 36 cancers in 185 countries. CA Cancer J Clin 2021; 71(3): 209-49.
[http://dx.doi.org/10.3322/caac.21660] [PMID: 33538338]

[10] Koelle K, Martin MA, Antia R, Lopman B, Dean NE. The changing epidemiology of SARS-CoV-2. Science 2022; 375(6585): 1116-21.
[http://dx.doi.org/10.1126/science.abm4915] [PMID: 35271324]

[11] Bojkova D, Klann K, Koch B, *et al.* Proteomics of SARS-CoV-2-infected host cells reveals therapy targets. Nature 2020; 583(7816): 469-72.
[http://dx.doi.org/10.1038/s41586-020-2332-7] [PMID: 32408336]

[12] Raman B, Bluemke DA, Lüscher TF, Neubauer S. Long COVID: post-acute sequelae of COVID-19 with a cardiovascular focus. Eur Heart J 2022; 43(11): 1157-72.
[http://dx.doi.org/10.1093/eurheartj/ehac031] [PMID: 35176758]

[13] Tang SW, Leonard BE, Helmeste DM. Long COVID, neuropsychiatric disorders, psychotropics, present and future. Acta Neuropsychiatr 2022; 34(3): 109-26.
[http://dx.doi.org/10.1017/neu.2022.6] [PMID: 35144718]

[14] Mirabelli C, Wotring JW, Zhang CJ, *et al.* Morphological cell profiling of SARS-CoV-2 infection identifies drug repurposing candidates for COVID-19. Proc Natl Acad Sci 2021; 118(36): e2105815118.
[http://dx.doi.org/10.1073/pnas.2105815118] [PMID: 34413211]

[15] Sonkar C, Doharey PK, Rathore AS, *et al.* Repurposing of gastric cancer drugs against COVID-19. Comput Biol Med 2021; 137: 104826.
[http://dx.doi.org/10.1016/j.compbiomed.2021.104826] [PMID: 34537409]

[16] Urbina F, Puhl AC, Ekins S. Recent advances in drug repurposing using machine learning. Curr Opin Chem Biol 2021; 65: 74-84.
[http://dx.doi.org/10.1016/j.cbpa.2021.06.001] [PMID: 34274565]

[17] Baalbaki N, Duijvelaar E, Said MM, *et al.* Pharmacokinetics and pharmacodynamics of imatinib for optimal drug repurposing from cancer to COVID-19. Eur J Pharm Sci 2023; 184: 106418.
[http://dx.doi.org/10.1016/j.ejps.2023.106418] [PMID: 36870577]

[18] Lucchetta M, Pellegrini M. Drug repositioning by merging active subnetworks validated in cancer and COVID-19. Sci Rep 2021; 11(1): 19839.
[http://dx.doi.org/10.1038/s41598-021-99399-2] [PMID: 34615934]

[19] Paul G, Mahajan RK, Mahajan R, Gautam P, Paul B. Systemic manifestations of COVID-19. J Anaesthesiol Clin Pharmacol 2020; 36(4): 435-42.
[http://dx.doi.org/10.4103/joacp.JOACP_359_20] [PMID: 33840920]

[20] Gowda P, Patrick S, Joshi SD, Kumawat RK, Sen E. Repurposing Methotrexate in Dampening SARS-CoV2-S1-Mediated IL6 Expression: Lessons Learnt from Lung Cancer. Inflammation 2022; 45(1): 172-9.
[http://dx.doi.org/10.1007/s10753-021-01536-6] [PMID: 34480250]

[21] Tesei A, Cortesi M, Bedeschi M, *et al.* Repurposing the antiplatelet agent ticlopidine to counteract the acute phase of er stress condition: An opportunity for fighting coronavirus infections and cancer. Molecules 2022; 27(14): 4327.
[http://dx.doi.org/10.3390/molecules27144327] [PMID: 35889200]

[22] Singh N, Baby D, Rajguru J, Patil P, Thakkannavar S, Pujari V. Inflammation and cancer. Ann Afr Med 2019; 18(3): 121-6.
[http://dx.doi.org/10.4103/aam.aam_56_18] [PMID: 31417011]

[23] Sawa T, Akaike T. What triggers inflammation in COVID-19? eLife 2022; 11: e76231.
[http://dx.doi.org/10.7554/eLife.76231] [PMID: 35049500]

[24] Madamsetty VS, Mohammadinejad R, Uzieliene I, *et al.* Dexamethasone: Insights into pharmacological aspects, therapeutic mechanisms, and delivery systems. ACS Biomater Sci Eng 2022; 8(5): 1763-90.
[http://dx.doi.org/10.1021/acsbiomaterials.2c00026] [PMID: 35439408]

[25] Krause T, Gerbershagen MU, Fiege M, Weißhorn R, Wappler F. Dantrolene : A review of its pharmacology, therapeutic use and new developments. Anaesthesia 2004; 59(4): 364-73.
[http://dx.doi.org/10.1111/j.1365-2044.2004.03658.x] [PMID: 15023108]

[26] Wei H, Liang G, Vera RM. Dantrolene repurposed to treat sepsis or septic shock and COVID-19 patients. Eur Rev Med Pharmaco 2021; 25(7): 3136-44.
[PMID: 33877683]

[27] Lechner-Scott J, Levy M, Hawkes C, Yeh A, Giovannoni G. Long COVID or post COVID-19 syndrome. Mult Scler Relat Disord 2021; 55: 103268.
[http://dx.doi.org/10.1016/j.msard.2021.103268] [PMID: 34601388]

[28] Kato Y, Nishiyama K, Nishimura A, *et al.* Drug repurposing for the treatment of COVID-19. J Pharmacol Sci 2022; 149(3): 108-14.
[http://dx.doi.org/10.1016/j.jphs.2022.04.007] [PMID: 35641023]

[29] Rao PPN, Pham AT, Shakeri A, *et al.* Drug repurposing: Dipeptidyl peptidase IV (DPP4) inhibitors as potential agents to treat SARS-CoV-2 (2019-nCoV) infection. Pharmaceuticals 2021; 14(1): 44.
[http://dx.doi.org/10.3390/ph14010044] [PMID: 33430081]

[30] Mohan S, Dharani J, Natarajan R, Nagarajan A. Molecular docking and identification of G-protei-
-coupled receptor 120 (GPR120) agonists as SARS COVID-19 MPro inhibitors. J Genet Eng Biotechnol 2022; 20(1): 108.
[http://dx.doi.org/10.1186/s43141-022-00375-8] [PMID: 35849279]

[31] Sanchez-Rangel E, Inzucchi SE. Metformin: Clinical use in type 2 diabetes. Diabetologia 2017; 60(9): 1586-93.
[http://dx.doi.org/10.1007/s00125-017-4336-x] [PMID: 28770321]

[32] LaMoia TE, Shulman GI. Cellular and molecular mechanisms of metformin action. Endocr Rev 2021; 42(1): 77-96.
[http://dx.doi.org/10.1210/endrev/bnaa023] [PMID: 32897388]

[33] Zhao B, Luo J, Yu T, Zhou L, Lv H, Shang P. Anticancer mechanisms of metformin: A review of the current evidence. Life Sci 2020; 254: 117717.
[http://dx.doi.org/10.1016/j.lfs.2020.117717] [PMID: 32339541]

[34] Nurcahyanti ADR, Jap A, Lady J, *et al.* Function of selected natural antidiabetic compounds with potential against cancer *via* modulation of the PI3K/AKT/mTOR cascade. Biomed Pharmacother 2021; 144: 112138.
[http://dx.doi.org/10.1016/j.biopha.2021.112138] [PMID: 34750026]

[35] Triggle CR, Mohammed I, Bshesh K, *et al.* Metformin: Is it a drug for all reasons and diseases? Metabolism 2022; 133: 155223.
[http://dx.doi.org/10.1016/j.metabol.2022.155223] [PMID: 35640743]

[36] Morale MG, Tamura RE, Rubio IGS. Metformin and cancer hallmarks: Molecular mechanisms in thyroid, prostate and head and neck cancer models. Biomolecules 2022; 12(3): 357.
[http://dx.doi.org/10.3390/biom12030357] [PMID: 35327549]

[37] Hashemi P, Pezeshki S. Repurposing metformin for covid-19 complications in patients with type 2 diabetes and insulin resistance. Immunopharmacol Immunotoxicol 2021; 43(3): 265-70.
[http://dx.doi.org/10.1080/08923973.2021.1925294] [PMID: 34057870]

[38] Matsuoka Y, Morimoto S, Fujishiro M, *et al.* Metformin repositioning in rheumatoid arthritis. Clin Exp Rheumatol 2021; 39(4): 763-8.
[http://dx.doi.org/10.55563/clinexprheumatol/zn2u9h] [PMID: 32828146]

[39] Salvatore T, Pafundi PC, Galiero R, *et al.* Metformin: A potential therapeutic tool for rheumatologists. Pharmaceuticals 2020; 13(9): 234.
[http://dx.doi.org/10.3390/ph13090234] [PMID: 32899806]

[40]   Ng L, Foo DCC, Wong CKH, Man ATK, Lo OSH, Law WL. Repurposing DPP-4 inhibitors for colorectal cancer: A retrospective and single center study. Cancers 2021; 13(14): 3588.
[http://dx.doi.org/10.3390/cancers13143588] [PMID: 34298800]

[41]   Hendriks AM, Schrijnders D, Kleefstra N, *et al.* Sulfonylurea derivatives and cancer, friend or foe? Eur J Pharmacol 2019; 861: 172598.
[http://dx.doi.org/10.1016/j.ejphar.2019.172598] [PMID: 31408647]

[42]   Allain H, Bentué-Ferrer D, Tribut O, Gauthier S, Michel BF, Rochelle CD-L. Alzheimer's disease: The pharmacological pathway. Fundam Clin Pharmacol 2003; 17(4): 419-28.
[http://dx.doi.org/10.1046/j.1472-8206.2003.00153.x] [PMID: 12914543]

[43]   Husna Ibrahim N, Yahaya MF, Mohamed W, Teoh SL, Hui CK, Kumar J. Pharmacotherapy of alzheimer's disease: Seeking clarity in a time of uncertainty. Front Pharmacol 2020; 11: 261.
[http://dx.doi.org/10.3389/fphar.2020.00261] [PMID: 32265696]

[44]   Mansour HM, Fawzy HM, El-Khatib AS, Khattab MM. Lapatinib ditosylate rescues memory impairment in D-galactose/ovariectomized rats: Potential repositioning of an anti-cancer drug for the treatment of Alzheimer's disease. Exp Neurol 2021; 341: 113697.
[http://dx.doi.org/10.1016/j.expneurol.2021.113697] [PMID: 33727095]

[45]   Mansour HM, Fawzy HM, El-Khatib AS, Khattab MM. Potential repositioning of anti-cancer EGFR inhibitors in alzheimer's disease: Current perspectives and challenging prospects. Neuroscience 2021; 469: 191-6.
[http://dx.doi.org/10.1016/j.neuroscience.2021.06.013] [PMID: 34139302]

[46]   Desai RJ, Varma VR, Gerhard T, *et al.* Comparative risk of alzheimer disease and related dementia among medicare beneficiaries with rheumatoid arthritis treated with targeted disease-modifying antirheumatic agents. JAMA Netw Open 2022; 5(4): e226567.
[http://dx.doi.org/10.1001/jamanetworkopen.2022.6567] [PMID: 35394510]

[47]   Sui S, Lv H. Cognitive improving actions of tofacitinib in a mouse model of Alzheimer disease involving TNF-α, IL-6, PI3K-Akt and GSK-3β signalling pathway. Int J Neurosci 2022; 1-9.
[http://dx.doi.org/10.1080/00207454.2022.2151712] [PMID: 36503352]

[48]   Zakaria S, Ansary A, Abdel-Hamid NM, El-Shishtawy MM. Dantrolene potentiates the antineoplastic effect of sorafenib in hepatocellular carcinoma *via* targeting Ca^{+2}/PI3K signaling pathway. Curr Mol Pharmacol 2021; 14(5): 900-13.
[http://dx.doi.org/10.2174/1874467214666210126110627] [PMID: 33573585]

[49]   Shi Y, Zhang L, Gao X, *et al.* Intranasal dantrolene as a disease modifying drug in alzheimer 5XFAD Mice. J Alzheimers Dis 2020; 76(4): 1375-89.
[http://dx.doi.org/10.3233/JAD-200227] [PMID: 32623395]

[50]   Wang Y, Liang G, Liang S, Mund R, Shi Y, Wei H. Dantrolene ameliorates impaired neurogenesis and synaptogenesis in induced pluripotent stem cell lines derived from patients with alzheimer's disease. Anesthesiology 2020; 132(5): 1062-79.
[http://dx.doi.org/10.1097/ALN.0000000000003224] [PMID: 32149777]

Recent Advances in Biotechnology, 2024, Vol. 9, 216-221

SUBJECT INDEX

A

Acid 10, 90, 94, 95, 96, 97, 100, 104, 106,
119, 122, 123, 125, 140, 142, 143, 144,
146, 147, 148, 150, 208
 acetic 146
 anacardic 123
 ascorbic 100, 147
 asiatic 122, 123
 caffeic 95, 96, 144
 chlorogenic 143, 144
 coumaric 142
 ellagic 104
 fatty 96, 150
 ferulic 96
 formic 146, 147
 gallic 96, 125, 144
 glucuronic 119
 hyaluronic 104
 lactic 146, 147
 linoleic 208
 organic 140, 142, 146, 147
 pachymic 10
 phenolic 90, 94, 95, 106, 140, 144
 propionic 146
 protocatechnic 144
 rosmarinic 96, 97, 123
 tetramic 148
Acidic fibrillary protein 211
Active thiocyanate 150
Activity 10, 24, 36, 53, 54, 55, 56, 57, 58, 65,
90, 96, 97, 98, 101, 102, 118, 122, 123,
126, 127, 128, 138, 142, 145, 151, 208,
210
 anti-inflammatory 118, 123
 antibiofilm 58, 128
 anticancer 90
 antifibrotic 118, 127
 antimicrobial 53, 118, 128, 138, 142, 145
 antiseptic 151
 antitumor 101, 102
 enzyme inhibition 98

 influence antibiotic 54
 inhibitory 24, 36
 photoprotective 96, 101
 proapoptotic 97
 protein tyrosine kinase 102
Adaptive immune systems 176
Adeno-associated viruses (AAVs) 70, 72, 74,
76, 79
Adenocarcinoma 208
Adenovirus vector vaccines 74
Agents 78, 99, 139, 143, 148, 153, 169
 antiviral 148
 chemopreventive 99
 immunological 78
 infectious 169
 natural antimicrobial 139, 143
 synthetic antimicrobial 139, 153
Alzheimer's disease 165, 173, 204, 205, 208,
210, 211
Anti-inflammatory 96, 99, 207, 211
 effects 99, 207, 211
 responses 96
Antibacterial 54, 57, 58, 64, 125, 128, 129,
137, 138, 140, 141, 142, 143, 144, 145,
146, 148, 149, 150, 151, 152, 154
 action 58, 142, 143, 150
 activity 57, 58, 125, 128, 140, 142, 145,
146, 148, 150, 151, 152, 154
 agents 58, 64, 128, 142, 148
 effects 58, 125, 129, 141, 144, 154
 efficacy 54, 137, 140, 142, 143
 properties 138, 140, 141, 142, 143, 144,
145, 146, 148, 149, 151, 154
Antibiotic(s) 53, 54, 55, 56, 57, 58, 59, 60, 61,
65, 66, 137, 138, 139, 145, 146, 152,
153, 154, 195, 196
 broad-spectrum 57
 natural 138, 139
 resistance 53, 54, 55, 59, 66, 145, 195, 196
Antibodies, monoclonal 176
Antimicrobial 64, 137, 138, 140, 142, 144,
145, 146, 148, 150, 152

www.ingramcontent.com/pod-product-compliance
Lightning Source LLC
Chambersburg PA
CBHW050835220326
41598CB00006B/372